INDIRECT

A Vision of Reality

A new theory of the universe

INDIRECT

A Vision of Reality

A new theory of the universe

Ernest J. Liebscher

Ontic Publishing

Book design and production by Tabby House

Manufactured in Canada

ISBN: 0-9681766-0-7

Cover design by Pearl & Associates

Ontic Publishing

Oshawa, Ontario L1H 7K4

Contents

Introduction

THE SEARCH FOR THE MEANING OF LIFE has occupied the mind of man since unknown times. In the later course of the search, when literacy spread, various grand models of the world have been projected. These models were relevant in their time and convinced many to be practical as guideline for their lives. When, in time, the supporting facts changed, it was difficult to revise the model because, once adopted, such models resist change and fight for self-preservation. From ancient times, the individuals who suggested and administered such models to the people were the magicians, shamans and seers. Then came the prophets, and later the philosophers, who interpreted the world for people who were too preoccupied to think for themselves. In more recent times, experimental scientists have taken this task on themselves and influenced philosophers and almost everyone else who thinks about the design of reality as merely mechanical, explainable by mathematics.

The influence of the specialized sciences is supported by independent discoveries, without commitment to integration into a meaningful whole and, without asking questions about the consequence of a discovery on the larger unity of the perceived reality. These models do not try to fit all the discoveries into a rational system. Instead, they imply that as long as something works, its usefulness is proven. This carefree attitude will have unforeseen consequences in the distant future.

Occasionally, a self-assured scientist has ventured to propose a model based on the physical world. Usually it is a one-sided mechanical arrangement derived from data in which conscious humans have no place. This denies us the experience of a meaningful existence in a totally material world that happened to exist by accident.

As far as the collection of provable data is concerned, we have progressed to the point of saturation. Additional discoveries about nature only increases the quantity of data, but does little to help us understand the underlying meaning. Since the sciences cannot understand the meaning of the world from collected data, they concluded there is no meaning. If one disagrees with this simplistic view, all one can do is disprove it by providing facts. But to prove them wrong is not enough. One must have something better to take its place. This I have done by searching for recurring patterns in nature to prove there is a purpose to reality. I found that these patterns also pointed to extrasensory phenomena in a theory of the meaning of life because we are going nowhere unless we are willing to accept that chaos and chance are also part of the essence of reality, and existence is meaningless.

In nature, the underlying presence of order and design in the higher evolved levels—the complex organization in living beings—point to order behind the apparent chaos. Of importance also is that we already know that all things have evolved from one common substance. Physicists call this substance "primal energy" and they presently are trying to discover the mechanism of nature through a formula that is supposed to explain everything that happens in the universe. This basic formula, pertaining to energy, they call the Universal Principle, or the Grand Unified Theory, which many believe explains everything. The idea of a common origin is correct, but I disagree with the assumption that it should be based on a mechanical process explained by a mathematical equation. This concept contradicts the presence of conscious beings, and ignores the importance of organization in nature and reality. The existing facts contradict a mathematical explanation in a mechanical formula. The reverse is more realistic—that conscious organization can explain the universe as meaningful. The physical is a secondary structure for the continuing evolution of consciousness.

Nevertheless, the possibility of the existence of a common principle is *the* major issue in our present time, to which psychology can contribute to explain consciousness and meaningful organization in the world. Philosophy has to reorganize many ideas to fit into new concept of reality. Many thinkers know that a new model is necessary to explain many recent discoveries, but we cannot expect new ideas from the physical sciences.

Even if we cannot expect to know more than the general principle of the universe, we should be able to organize a provisional model that includes rational beings as integral. I think this point has been reached in our present time, as sufficient knowledge begins to merge and culminate into a common principle for all that exists.

Even if nuclear physics was first to perceive the electrical basis of the material structures, the understanding of the whole principle is imminent now. If it should turn out that consciousness and electromagnetic energy has a common origin, that nature has evolved and organized by conscious orientation to meaning, it has certainly never been by blind mechanisms. Something good has come from the skeptical interlude of the last two hundred years of science, when most of the fixed ideas of the past have been stripped away. We could now put a new conceptual model of the world together again and it is with a different understanding of what is and is not possible.

This book is derived from many ideas and sources, positive and negative, to achieve a synthesis and reduce everything to a single, common principle. In our thirst for knowledge we would not have rested anyway, until we found the key to understand the meaning of being. Sooner or later we would have found out what makes the world run.

This book is about the Principle of Reality in a general sense, aware that the details are constantly changing in an open-ended process. The first chapter challenges the presently established concept of reality. The following chapters outline a new model of reality from which the larger Universal Principle is developed. The principle is expanded by comparing patterns to existing facts. Because some speculation is involved to connect ideas that fall outside of physical prove, there are metaphysical parallels presumed, but my theory is based on facts in the reality of the world.

1

About Reality

IT IS IRONIC THAT IN OUR TIME, when technology has improved human life-styles as never before, when computers allow the processing of information and the spread of knowledge worldwide, and when inventors are experimenting to create artificial intelligence, that we should begin to doubt whether or not we are headed in the right direction. Instead of feeling elated, some thoughtful people worry—not just about the side effects of this great progress from industrialization, but also that the general perception of the meaning of existence and the universe must be profoundly wrong.

The mechanization that is proliferating everywhere is ignoring living nature and conscious beings, and this has caused a wave of opposition that is mostly expressed as a return to nature. It has also caused philosophers to reconsider where we have taken the wrong direction—mechanization—in our thinking about the meaning of reality. This controversy has raised the possibility that the opposite is more likely true—that instead of the mechanical universe of scientific materialism, the universe is a conscious process, and everything in it has a psyche which has evolved from a common essence into the multiplicity of the countless psychic entities that are all a part of a greater psyche pervading the universe.

That we humans have a separate psyche in the age of psychology is already accepted. That other beings of the vast universe should also have some form of consciousness can be assumed. We just have to find the connection where all this fits together with the mechanical universe of physics.

If philosophy and psychology could come up with a concept to explain the world as a conscious process, it would be rejected by experimental scientists as speculation since they are convinced they have a sure method to establish the truth. This method, based on physical verification, is where the controversy always starts. How can one prove the existence of consciousness when it is not a physical object?

Science has found an explanation, namely that consciousness is just a side effect of the physical nature of materials, which is real. Consciousness is merely a form of mental referee without relevance to the real world, just imagery of a later-evolved mist that rose as evaporation. This irrelevant mist was the mystical past of mankind, the occult imagery that made early humans believe in capricious gods and became later reduced in most cultures to one God. From this idea theology evolved, which then turned into philosophy that ended up with a spiritual explanation of reality. It was from philosophy that skeptics began to question the wild speculation of human imagination that presumed that behind the physical world is a larger metaphysical reality which became other-worldly, namely spiritualism.

Eventually the separation from philosophy came in the sixteenth century, during the Renaissance, when empirical methods were introduced that based reality on physical terms. This was the empirical method of measurement and quantification. With this step science became independent of speculation and based truth on physical proof. Science also adopted mathematical verification as more reliable than speculation. Materialism thereby became the foundation of science, conscious thinking became misleading imagery or wishful reasoning. Nature was explained as a mechanical process following laws established by physics. Everything is mechanical and science is discovering the mechanisms by which all things, and the universe as a whole, work. This will be the mechanical principle of the universe. In fact, that is exactly what experimental science currently is perusing, only it has sightly reduced the claim from the universal principle of everything to the principle of energy from which everything is derived. This has happened because scientists have found out that they are conscious beings and they also use their imagination to speculate. This means that without consciousness this mechanical system is unreal. Therefore, they added conscious beings as a freak accident to their physical universe and this is why they cannot claim to search for a universal principle "without exception," unless they want to ignore billions conscious beings as irrelevant—including themselves.

This ends the single principle with a major flaw that they would have to reject themselves. Common sense tells us that the world is the total opposite from mechanical, namely a conscious process that has evolved a meaningful world where everything has some purpose for its existence—to give it meaning for being. The simplest proof against accidental happening is the presence of organization everywhere. In fact, our bodies are organized from atoms and molecules which in turn are organized into cells, and from cells the organization has gradually expanded into mobile self-conscious beings of which we humans are a stage, but not necessarily the highest stage or the last level.

While these intellectual opinions have been debated between science and philosophy, in the practical world mechanization has progressed through ideas. For example, inventors have discovered electricity, which has proven to be invisible energy without material substance. Eventually the physical atom was exposed as purely immaterial, consisting only of waves, or particles of waves, vibrating as separate entities without any substance. This should have ended materialism, but it did not, since the vibration that produced energy, even without content, could be called physical because vibrations could be measured and calculated. Physics survived by calling energy a sub-form of matter.

Physics carried on with the old model of the material world when reality could have been explained as an organization of waves. But admitting organization would have implied that the entire mechanical concept of physics based on accidental evolution is wrong, since organization is the opposite of accidental evolution. The fact is that the majority of experimental scientists remained loyal to the material model of the tangible reality.

Perhaps organization had not even occurred to them as important since their minds were programmed for mathematics based on the universal law of physical gravitation. This law, discovered by Isaac Newton, explained the universe as a mechanical clockwork based on gravity. When, two hundred years later, a new theory of wave mechanics, now called the Quantum Electro-Dynamics Theory, was developed that could dispense with gravity, the majority again remained faithful to materialism and universal gravitation.

The conservative majority of scientists still believe in materialism and the mechanical universe, even resorting to calling philosophy and psychology useless imagery that at best is misleading science and has never done anything, particularly when physics provided complex mathematical equations to explain nuclear energy as the reason for the formation of the universe.

The trust in universal gravitation has currently reached the limits of credibility. According to the Grand Unified Theory of physics, some theoretical physicists are close to combining the Electro-Dynamic Theory with the Theory of Gravity to formulate the final mathematical equation of the principle of the universe. According to this theory the universe is supposed to have come into existence with a "Big Bang." Based on the presumed existence of gravity which is the "ultimate force," the force of gravity is supposed to have compressed all matter in universe into a single ball that then exploded, and from this moment time began.

This is the wildest theory of physics ever to come out, and especially in the face of the fact that in the universe there is no gravitational force influencing the heavenly bodies, which is evident from space flight. Once a spacecraft is outside the limits of the field of gravity that holds objects to the ground, astronauts float freely about and objects travel in a straight line. The only purpose of gravity on Earth is to prevent things from falling into space. That is all.

The origin of gravity

The idea of gravity is actually based on experiments made by Galileo, a professor of mathematics and mechanics who lived in Italy four hundred years ago, and who had developed a formula for free fall, which he assumed to be caused by magnetism. And he assumed that magnetism held the moon in a perfect distance from the Earth, equal to the centrifugal and centripetal force as a satellite of our planet. Galileo also assumed that the planets are satellites of the sun traveling in perfect circles according to their magnetic properties.

A generation later Isaac Newton expanded this formula into a heavenly mechanics, but had to "invent gravity" because magnetism could be blocked out. Yet nobody has been able to confirm that gravity is a physical force. It is an idea necessary to explain an effect for which physical evidence does not exist. That means that the idea of gravity misled physics and cosmology for three hundred years. Even the quantum theory could not discourage it. But Newton was only following Galileo; it was Galileo who misled physics. And he did it despite the information provided him by the mathematician, Johannes Kepler, who wrote to him explaining from exact measurements that the planets follow elliptical orbits. This information Galileo regarded as an error, because that would have meant that mechanical magnetism was a wrong idea. In fact, Galileo is still correct that mechanical satellites require a circular orbit. Obviously planets are not mechanical contraptions but naturally evolved bodies that pursue their paths intentionally, correcting them occasionally.

Now in our time the idea of gravity expanded by Newton has been taken to the extreme in absurdity by a mathematics developed by Einstein, Minkowsky and Riemann, making everything relative. Although Einstein was correct in certain details of physics, as was Newton, with regard to universal gravitation they were wrong. If gravity is the ultimate force, this assumption, already mentioned, was taken to the extreme by other physicists, thinking that gravity compressed all matter of the entire universe into a single ball which eventually exploded with a big bang creating extremely high temperatures. Since then all matter has been drifting apart, which physicists now explain as the red shift because galaxies seem to be distancing themselves from each other.

Nuclear physicists have experimented with atomic energy trying to fuse atoms together, but they have never succeeded in over fifty years. The reason why they think it is impossible is only because on the planet Earth we cannot recreate temperatures high enough to fuse atoms. Even the temperature in the sun is not high enough to fuse the complex atoms that exist on Earth. Therefore, they think that such temperatures must have occurred in the first second of the big bang explosion.

Governments have spent billions of dollars working on nuclear fusion and still have achieved no results. The fundamental error made by experimental scientists in general was to assume that nature used the same methods to create things as they use in fusing chemicals, namely heat. That is a major misunderstanding since nature is unlikely to have employed the same methods scientists use in laboratory experiments. Nature has had immeasurable time available, and we can see that atoms are organized into meaningful patterns and posses qualities of their own, not just quantity. Quantification is missing quality and meaning, which is the most important aspect in human life. Without meaning and quality, life is worthless. This is another major misunderstanding that occurred when science separated from philosophy.

The symmetric arrangement of atoms is revealing a meaningful pattern of order that started from the first wave and organized into particles and atoms, which again organized into a great variety of worlds with different qualities. Physics obviously is a one-sided perspective of reality that is unlikely to explain reality because it is completely based on materialism without meaning.

Not only is the "law of universal gravity" a myth since there is no gravity in space, the reliance on experimental proof is a narrow method that rejects human experience as subjective. Looking back into human history one will find that this so-called scientific age began in that period known as the Renaissance when human knowledge made a general step forward.

When did this self-deception, which is called the scientific method, start and make its way into the higher institutions of learning and research? We know that Galileo was promoting the solar system of Copernicus, who claimed that the sun is the center around which the planets circle. By improving the telescope Galileo pointed out that the moon is a satellite of the planet Earth, held in balance by equal centrifugal and centripetal force according to its mass.

Galileo never actually took measurements of the planets that were circling the sun, and that was where the mistake started. Galileo, then professor of mechanics in Padua, Italy's foremost university, had received and ignored that critically important letter from Kepler, who lived in an obscure place called Prague, thereby making a grand error which we only now have found out by space travel. This means that all subsequent theories of mechanical gravity and mass are wrong, as is the entire celestial mechanics of Newton and all that is based on mechanical gravity.

Let us now find out who the real scientists were, those who actually calculated the true orbits of the planets on which modern spaceflight is based. Although the University at Padua in Italy was then the best-known institution of learning in Europe, there were persons who seriously disagreed with the dominance of theology over science and philosophy. One such person was the Emperor Rudolf II, who had set up a protective haven for unorthodox freethinkers in his castle in Prague. This happened while the Roman Church was persecuting people everywhere else who disagreed with the dogma of the church. Why could the Holy Inquisition not force Rudolf to stay within the holy dogma? The irony was that the Emperor, who was supposed to protect Christendom, indirectly helped to undermine it. The church was powerless because the Holy Roman Emperor, Rudolf II, was equal to the pope in status. He was supposed to be the military arm of the papacy. Many things were not as they should have been, because during the Renaissance belief in the authority of the church was declining. Even among the faithful, the belief in the principle of Christianity was seen as violated by the upper caste of the church, who had become just as materialistic as the worldly feudalistic aristocracy, and were exploiting their subjects just as badly. The call for reformation of the church had already resulted in the spread of fundamentalist sects known as Protestants before Rudolf became kaiser (caesar). This situation existed as a deep split in the population, while in the south the Turks invaded Hungary, which was also part of the Holy Roman Empire. All of these situations demanded diplomacy and concessions to many sides, a difficult balance for any human.

The emperor was the first son of Kaiser Maximillian II, of the ruling Hapsburg family. The Hapsburgs, who were originally German, had intermarried with other ruling families of Europe, in particularly with those of Spain. Rudolf grew up in Vienna and was educated in Spain, which at that time was colonizing the pagans in the Americas. Obviously Rudolf was better informed than most at that time. While residing in Prague, his interest in science and the arts caused him to support more than two hundred intellectuals from all corners of Europe who were living under his protection. Most of these Latin-speaking intellectuals, then called alchemists, believed in unknown powers and mysterious forces in nature. They searched for the "stone of wisdom," similar to what physicists are still searching for today—the principle of transformation of substances, and not exclusively through chemistry, since the alchemists were also open-minded to mysterious powers, innate in substances, that could be influenced by magic.

This kind of thinking was apparently shared by Rudolf, who personally associated with the alchemists to discuss such questions. This behavior was considered inappropriate for a protector of the Catholic Church to whom the support of alchemists was an annoyance. To the aristocracy, he was far too friendly with commoners, and to the pious people he was described an agnostic, yet he also had loyal admirers among the people. Rudolf managed to hold the imperial crown for more than thirty years before he was forced out as being indecisive, mostly because he tolerated the right of other beliefs to exist. Rudolf's indecision to act was actually wisdom: He left religion alone. He was replaced by his younger brother. The new kaiser complied with the pope's wish to enforce Catholicism in a "counter reformation," which soon developed into a fanatic religious war that lasted thirty years, decimating the population and dividing the empire into many principalities and kingdoms. The church did not win this cruel and useless war that it had started; nobody won. The war ended with the exhaustion of the lands and the people. In a declaration of peace it was decreed that religion should be a personal belief and that the state should separate itself from religion. With this act the de facto Holy Roman Empire broke up. German and Austrian kings still were addressed as kaiser.

Let me return to the actual subject of why the alchemists of Prague were more conscientious in their research than the professors at the University of Padua, which was nearer to Rome. In Prague the religious atmosphere was tolerant. Protestants, Hussites and Jews mixed freely with Catholics. Witchcraft and other mysterious phenomena, which in our time are called extrasensory experiences, were considered to be possible, and psychokinesis was considered magic. In fact, Rudolf himself was interested in such phenomena but, in general, people were careful about the opinions they held and to whom they expressed them, because in Padua the philosopher Giordano Bruno had been burned at the stake for spreading agnostic ideas.

Rudolf also wanted to know whether or not Copernicus was correct and he invited the astronomer Tycho Brahe to Prague. Tycho had made accurate measurements of the planets and collected reliable data, but could not confirm the theory of Copernicus, that the sun was the center of the solar system. Nor could any of the others who had also made actual measurements. If there was a center, it lay somewhere beside the sun. The principle was convincing but the details were not.

Tycho hired the mathematician Kepler as his assistant. After Tycho's sudden death, Kepler took over the project and from the measurements they had collected, tried many different approaches to solving the puzzle. Because the planets are in constant motion relative to each other, and the Earth is turning on its axis as well, the exact time when the measurements were taken was a critical factor. This involved four-dimensional dynamics well before Einstein came to call time the fourth dimension.

Kepler eventually isolated the planet Mars and concentrated on its path. He understood by this time that every planet had a different elliptical orbit. Kepler then developed a formula with which he calculated the Rudolfian Tables of the five known planets. These tables served astronomy for the next three hundred years, until the development of better instruments improved accuracy for the use of space flight in our time. Kepler also calculated astrological charts for high ranking persons, such as Count Wallenstein whose violent death he predicted. Kepler was open-minded. His views on reality were not necessarily similar to those of the alchemists and mystics. To make a comparison with Galileo reveals a difference in objectivity which had great consequences for the future of science.

Perhaps it was partly the dominance of theology in Padua which restricted Galileo to a mechanical concept. To theology the world was created by divine fiat, which meant that Galileo was limited to a mechanical interpretation. Newton, a generation later, also stayed clear of religious involvement and made God the mover of the mechanical clockwork universe based on universal gravitation. There are indications that Newton privately leaned towards alchemy but considered it wise to keep those ideas to himself.

The rejection of all extrasensory phenomena and spiritualism by science could still be an unconscious fear to avoid persecution. Scientists still avoid taking a position, even if the facts are clearly pointing to emotional causes and intentionality in nature. But now it may be intimidation by academic correctness that influences free expression. The experimental method is still a narrow mechanical concept that avoids all references to consciousness and spirit because these issues were once reserved for theology and philosophy. Obviously, all beings in nature have a purpose which they unconsciously or instinctively follow as their identity. This is a realistic perspective of living nature, not based on mysticism but consistent with behavior and actions with proof that organisms adjust themselves to conditions. Although members of lower species sometimes act as though they are programmed, they are not mechanical robots since they have self-orientation and can sense danger.

Around 1920, radio technology changed some ideas in a few open-minded physicists, causing them to abandon the concept of gravity and to develop a wave theory to meet the reality of new inventions in telecommunication technology. This new idea was called the theory of Quantum Wave Mechanics. Although it was another mechanical method to describe electronic waves with contrived mathematics, the development became mysterious when the presence of innate consciousness in subatomic particles pointed to "uncertainty" in their calculation. Not many wanted to believe it. One can therefore understand why the mathematical explanation became extremely complex and incomprehensible. But because it works with reality, the quantum theory eventually was accepted.

If reality is a wave phenomenon, then the waves that created the world are organized, which they obviously are, and at least they have a vague intention to evolve through organization. This intention must be assumed to be the "first cause" of the world. Elementary organization requires consciousness, which again implies that conscious orientation should exist in the void before meaningful worlds could evolve through organization. Total nothingness would never have organized the diversity of nature, with such complex organisms as we humans who have reached a level of self-consciousness and an understanding of the principles of nature. Something like innate consciousness must exist, at least in an elemental state from which the phenomenon of reality has evolved.

With the general deduction of meaningful organization in the real world, we can begin to speculate about the single principle that many rational persons expect must exist and is innate already in the primal substance from which everything has evolved. Therefore, the presence of organization which repeats itself in a multiplicity of expressions in "higher-evolved beings" has to be the starting point of this philosophical theory of reality. We can learn the most from a general view from the top down, from the most evolved and conscious, because they reveal more than the elementary particles.

Let us again look at some of the organized patterns of nature. We humans are conscious of many more interesting things than the animals are aware of. We have vastly more knowledge of the extent of the world than other beings and things that live in much narrower worlds, living their lives as they find them. We have expanded and enriched our perception of reality with a greater imagination; therefore our lives become more interesting, with the development of art and culture adding a refined dimension.

Besides the needs of food and shelter in order to survive, which all humans have in common, we are emotional beings in a meaningful world. From this perspective, mechanical theories are unrealistic. We live in cultural circles with different languages and particular symbolism, but the ultimate differentiation that separates each of us is individual identity. We are acutely aware that our identity is unique and our consciousness is nontransferable. When consciousness leaves, the body is dead. Instead of being unimportant as materialists claim, consciousness is the most important essence of existence. Without consciousness, nothing is known, but the reverse, that consciousness without a body is still conscious and could exist, is possible. Not just from a philosophical point but even more from actual experiences reported by humans who have had "out-of-body experiences" comes a long line of reports that are called subjective experiences of countless individuals who usually do not want to talk about them for fear of being ridiculed. The most important essence of life is lived experience that is outside of physical proof. In recent times some scientists have admitted that the presence of unconscious ideas that might organize chemicals or waves into meaningful designs is thinkable, but outside their field. As far as they are concerned this is metaphysics, which is philosophy.

In the progression of organization of nature, from waves to atoms, to cells and conscious humans, the pattern stretches from amorphous substances existing in large quantity towards complex organisms of specific quality, and fewer in number. Thus a pattern of increasing organization is clearly displayed in physical nature and continues on the mental level of organization of thought resulting in conscious ideas. This later progression is experienced as perception which is the ability to organize impressions into meaning and gestalt. Although the average, normal individual has only a limited horizon of direct interests and lives with fragments of knowledge, a few interested individuals organize the fragments into patterns that make sense as a whole in a greater nature. These individuals are not the only official philosophers—everyone can inform himself and organize things in different ways.

Obviously there are many explanations of reality possible, not just whether other species are conscious but whether their design is limited. We humans have five external senses plus imagination, which allows us to make sense of our perceptions by organizing them by meaning. The meanings we give to things come from our own personal ideas—ideas which have changed and improved constantly during our lifetime and in history, providing we are flexible and open minded.

Human beings have speculated about reality for thousands of years and their ideas have changed, but it is still through the mind that we organize images and attach meaning. Early man thought that other beings who shared our world might have ideas similar to humans, but we have found out this is farfetched. Although other species have meaning for themselves, their different designs demand other interests that have little in common with us. We seem to share only the basic level of emotion. This leaves the emotional connection as a possibility for transcendence of the separations between different species open, but it is still restricted to feelings. Therefore, a common principle that applies to all species would have to be emotional and psychological in essence.

For example, one night I drove to my home, which is situated in the country. When I turned off the headlights and walked towards the house, it was dark. I was momentarily startled when a "large being" walked towards me. I soon realized it was a horse. What does a horse feel on a cool, dark night? Does it feel alone and perhaps lost? I assume the horse came to me on that dark, lonely night as another living being. After it ran away from the barn it may have found out that freedom is only an idea, and changed its mind, realizing that a warm barn provided by humans is preferable to freedom. All I could do was speak to it as if to another human being, knowing that it can only sense feelings in my voice, hoping that telepathy may add to this effort.

In ancient Greece where western philosophy has its roots, realistic observations of nature were made that are still valid today. Socrates observed the presence of ideas in all things and beings, ideas that give them meaning and identity which is determined by their design and function. Socrates also concluded that ideas can exist as thought form; ideas that have social and moral meanings of good and evil, of justice and beauty; ideas that include emotions of love and compassion are meaningful. Abstract ideas, in particular justice and beauty, have an emotional basis that can transfer itself and can excite us, make us feel happy, angry or distraught. If such emotions are transferred by speech and gestures as ideas to a mass of people, they can mutually excite each other. Sometimes this can lead to violent acts. This is how emotions affect us. It is the way that wars which laid waste to entire regions have been started. It is like starting an emotional fire which spreads and consumes reason in a frenzy of insanity, because we lose control of reason.

Yet from the same emotional basis people can be moved to tears, to compassion and self-sacrifice, to deep love and loyalty, particularly in religious experiences. It depends on the ideas and values that inspire them. In fact, emotional orators can transfer their emotions to an audience and excite it into whatever mood they may wish to create, at least for a while, until the people in the audience regain their common sense and their ability to think for themselves when they are alone.

Because of the emotional nature of humans, intelligent persons understand the implications and are trying to cultivate the good side in people by appealing to their compassion and tolerance. Many humans are trying to civilize people into becoming socially responsible persons by advising control of their emotions. They have met with limited success. The problem is that people are always born with an impulsive nature. Starting from basic emotions, they have to learn to live in civilization. Progress is slow and seems futile when one sees how ignorance repeats itself in each generation.

From observation of nature, we can deduct that ideas are important for the identification of the meaning of things and beings, at least for us. What they are for themselves is unknown. While ideas are a rough guide to the "general" understanding of perceived reality, particular details are still open and subject to change. Modern psychology has found that the innate drive to assert oneself and expand can be slowed down and civilized, but it cannot be broken. If the spirit is broken, then it gives up and the organization becomes dysfunctional. The spirit that keeps the organization viable is not limitless in endurance; it is sensitive and vulnerable to negative experiences and can lose hope and the will to live and give up.

On the human level it is "life spirit" which inspires each being and is the essence of the identity. It is also that which holds the organization together. On other levels where we are not living and are not participating, one cannot know what others feel. But if we are all of the same essence, it is the same spirit that holds the atoms of the living and material things together as their identity, not mechanical gravity. It is also the spirit of identity to a purpose that keeps the planets performing their assigned roles, loyal to their mother star in the greater design of a meaningful universe.

Many thinking persons are opposed to mysticism and the occult, because one can be misled by taking them at face value. This is a serious misunderstanding—the separation between levels is great, apparently time and space are irrelevant in that other dimension and the meanings are expressed in symbols and opposites. One cannot take anything directly that comes from the other side. One cannot believe anything without analyzing it carefully. Regardless of what objections one has against mysterious phenomena, whether or not one has made up one's mind that they are all humbug, mysterious events actually do occur. Neither science nor religion can change the fact that there are unexplained phenomena. Certainly, denying them will not make the phenomena go away. By trying to explain them, we might progress further.

What I am aiming to achieve is the integration of all phenomena into a larger concept of reality by reorganizing everything that exists in any form—physical or thought—to make sense. A wave model allows the separation and coexistence of all phenomena into a single process without the need of categorically rejecting any experience as impossible.

Examples of extrasensory phenomena

The argument between science and philosophy is already in the past and not worth dwelling on, because in the next chapter a new model of reality will be introduced which is outside narrow physical verification, but not outside human experience through consciousness. Looking at our present civilization, we are deeply involved and dependent on mechanical inventions from technology and science. Without mechanization our civilization would face a catastrophe: therefore, we have to give them due consideration, but not blind trust. In fact, we cannot believe anything. The physical model is wrong about nature, and extrasensory perception can be equally misleading, only in a different way. All we have left is reason to sort things out.

For example, in ancient times fire was believed to be the substance of which the world was born. Physicists also think this is correct in principle, even if they now think it was nuclear fire. We know that fire means light and warmth and we humans in the organic mode also know that fire will cause pain if touched. We certainly know that we could not exist in fire, that it means certain death, whether chemical or nuclear. However, such physical facts, it has turned out, are not absolutely firm. In unusual states of mind one can become immune to fire, but to a limited degree only. Certainly, one could not survive inside a furnace, even if fire, like everything else, were illusion as the Hindus claim.

In the interest of reality, some outdated presumptions about physical reality have to be challenged with actual examples from human experiences and also by repeatable experiments, whenever possible, to satisfy science.

One can not deny the following examples because they are real. One example is fire-walking. It is uncommon and dangerous, but that it is possible at all is a serious challenge for the established concept of physical reality. There are other examples which defy common sense and have divided the opinions of people as long as they have been known. One is psychokinesis and another psychic healing, both of which were connected to magic. But that was long ago.

The first two examples I have chosen were reported when Europeans became interested in Asian religions and culture and travelled there to see the wonders first hand. These curious travellers were deeply impressed by the mental and physical capabilities of yogis and fakirs. The majority of the people in India, even today, are convinced that making vows, casting spells, praying and astrology actually can work if earnestly pursued. In principle this is similar to beliefs held by early Europeans before scientists supposedly proved such things impossible and called them plain superstition.

The art of fire-walking is still practiced in Sri Lanka and in carnival shows that appeared in the West. The idea that strong "belief" or hypnotic trance can make people immune to getting burned is certainly false since one gets burned regardless.

A story, obviously true, tells of a missionary from Europe who visited Sri Lanka around the beginning of this century, when Sri Lanka was British and still called Ceylon. He felt religiously challenged to find out whether or not his Christian belief would prove itself equal to the belief of the native Hindus and would also be able to protect him from injury. He therefore mustered the courage and ventured to take a few steps onto the bed of red glowing coals where he immediately collapsed and had to be pulled off quickly. He was taken to a British hospital with severe burns to his feet, legs and more, and spent half a year in the hospital. This, then, was supposed to have proved that Christian belief is inferior. The average Hindu, however, cannot walk through fire either, so something other than belief must be involved. In fact, many Hindus no longer think that religion plays a role, but that something entirely different is behind it.

The common view now is that a certain state of emotional detachment, of fearlessness and unreality about fire and pain is required, yet a hypnotic trance can only suspend the feeling of pain, not prevent the burning of flesh. Hypnosis would only suggest a false security which leaves the conclusion that deeper levels are involved. There could be a level on which the mind transcends reality or at least shields the human body from heat, but what it is we do not know exactly. The advice is, do not try it because you will get seriously burned. Besides, there is a limit to what organic beings can survive and a limit to transcendence in general. It should be made perfectly clear that the heat from the bed of red-hot coals is so intense that spectators keep a safe distance from it as clothes will ignite and hair singe just by standing near. Yet certain individuals in a "special" state of mind repeatedly walk barefoot back and forth over the long bed of glowing, red-hot coals without injury to their feet, clothes or even to a single hair. One interesting clue was given when during a demonstration in London when a fakir asked a reporter to come with him for a walk over the coals with the assurance that he would protect him. By holding on to the fakir's hand the reporter walked unhurt across the red coals. From this I conclude that a shield is extended around the fire walkers which protects them from physical reality.

Visitors to India have also reported that fakirs could perform a certain rope trick. The fakir sat on the ground in the open, in front of an expectant audience and began to play a melody on a flute to attract attention to himself. He then opened a basket in front of him. Out of that basket a rope slowly rose into the air as if by magic. A small boy appeared out of nowhere and started to climb the unsupported rope and disappeared into nothing. Informed spectators from Europe were prepared for this in advance and photographed the performance. The pictures, when later developed, showed only the fakir sitting on the ground with a basket, playing a flute.

To the spectators the experience seemed like a dream which they know is impossible even though their eyes see it. They know it was illusion, created in their minds by a fakir (or faker), and that settled the issue. Or does it really? To call this performance a trick played on the human mind is to avoid the implications. If we can be influenced to see mirages as groups, the implications about our perception of reality would be enormous. The consolation is, in order to achieve thought transference to such an extent, this is probably a gift one is born with. We must assume that the illusion was created by telepathic suggestion, and in fact, involved mass-suggestion.

Just for the record, telepathically gifted salesmen can create visions in the minds of prospective customers, so that the people themselves are convinced it was their own idea to buy things they did not need. Telepathic thought transference has long been known and is common among animals who do not speak but seem to sense what others think. Animals that are particularly telepathic to human thoughts are dolphins and orcas. Yet in the form of a public spectacle, telepathy is difficult to accept and it is scary to think that one can be so obviously deceived.

Thought transference is more common in ordinary life than one would believe, mainly among persons who are tuned into each other's thoughts. Often we humans have the same idea and know exactly what we mean when we talk face-to-face. That is why meeting with a person directly helps in understanding what he means.

We know that emotional orators can transfer their thoughts through emotions and create visual images in listeners' minds which occasionally has resulted in hysteria. One is reminded that in popular musical performances of singers, such as the Beatles or Elvis Presley it is not just the artistic performance but also the company of others that raises the mutual excitement to the level of hysteria.

From the example of the fakir we understand that he attracted and focused the attention by playing a flute. Obviously one must pay attention and thereby one opens a channel to telepathic thought transference. If one does not wish to be influenced, one is free to think of something else and remain unaffected. But if one came to see a show, one is already conditioned to expect something.

In our present time, certain gurus from India have moved to the industrialized West, to gather followers to contribute their earnings to the cause of a guru. That such gurus do have telepathic powers is undeniable, but one should know it is considered unethical to use them for material gains. Many followers of such cults claim to have seen divine proof in the form of light around their guru or a halo over the head. Some members have heard them speak inside their minds. But think again, a true yogi would not care for material wealth—he would not even consider exploiting the ignorance of people, but focuses his mind on joining Brahman, who is disinterested in the greedy world.

In a hypnotic trance, hallucinations can also be created by verbal suggestion, yet telepathic suggestion as thought is much more difficult to achieve. In the course of human history many charismatic prophets have appeared, and countless cults have flourished and passed away. People must have unconsciously experienced thought transference, not just from religiously inspired but also politically oriented persons who believed they had a mission to fulfil. Emotional orators like Cicero, or in our time, Hitler, could excite the masses and convince otherwise intelligent people to trust their judgement—a judgement which eventually had catastrophic consequences on the lives of millions. Charismatic persons are not necessarily farsighted or fair-minded individuals. Most are unconsciously driven by ideas which make them over confident and blind to reason.

Cult leaders often consider life of little value and have occasionally incited their followers to commit mass suicide, as in

Jonestown, Guyana, when all of the followers of Jim Jones took poison. One should remain skeptical to claims of divinity made by any human. Delusions of grandeur are common and difficult to recognize because such persons are usually intelligent and telepathic and seem normal. Incidently, transcendence is near the borderline between sanity and delusion. Besides, the entire subject of normality is uncertain at best. In this particular field, therefore, a clear mind is necessary since there is a difference in the consequences that follow.

Psychokinesis

Even in a new perspective of reality it is necessary that some physical proof must be evident, since a common principle also applies to the physical level. Many ideas that are physically expressed have metaphysical causes, not necessarily directly provable, which should make the controversial subject of "mind over matter" an interesting subject. Take for instance, the well-known case of Ted Cerios, a man of our time, who used to live in New York City. He amazed many reasonable persons by demonstrating the ability to create pictures by concentrating his mind on the lens of a camera (usually an Instamatic) which was loaded with unexposed film. While he mentally concentrated on any known landmark or object that people requested, the projected image of his thoughts appeared on the developed film. The pictures often lacked finer detail which is understandable, but the idea was there on the developed picture.[1]

This and similar phenomena are called "mind over matter" and were considered to be odd curiosities, given through the explanation that "exceptions prove the rule." Of course, the opposite is also true. Exceptions show that the entire concept of materialism is wrong and the mind can organize matter. The many exceptions and paradoxes that exist in reality prove that materialism is a limited concept, insufficient to explain many aspects of reality. Thought projection further confirms that mind is in control over matter and can influence other minds, within limits.

Thought projection actually implies something very important, namely that ideas can influence material substances and that the chemicals on the film organized into the idea that Cerios had in his mind. Regardless of how insignificant the effect may seem and that the quality of the pictures is inferior, it is important to know that thought can influence the organization of material substances and not just human minds. Just the fact that ideas can manifest in matter at all is a proof that ideas could be the cause of the organization of nature and everything that exists. Besides these deliberate experiments with mind over matter, other unintended cases have been reported such as when persons that should not have been in the picture were discovered on photographs. Sometimes faces of deceased persons are present on a photograph among the faces of the living. The latest case was when a mother happened to photograph her child who was sitting next to a television set that was turned off. The developed film showed the face of a human horizontal on the screen. Besides the photograph, this same television set sometimes had mysteriously turned itself on.

Such mysterious tales are many, but I cannot go into the vast collection of anecdotes—the principle can be found in a few example already. As the most important proof of the existence of consciousness in materials is Kirlian photography, invented in Russia. It shows the images of severed and disposed limbs still in their normal place when physically they do not exist. Does this mean that the ideas of things as they once were survive in a different electrical mode and can be recorded with a special technology? Besides this, persons who have lost a limb often report feelings as if the limb was still there. Most important is that human emotions are recorded as changes of color, proving that we are emotional. Even our health can be read by such colors.

A few special individuals have recently demonstrated that their thoughts can move material objects around. A Russian woman, Nina Kulagina, demonstrated that by concentrating her mind on material objects she could make them move a few centimeters.[2] Another Russian, a man by the name of Messing, could cause electric light bulbs to light up with his mind from a distance. Then there is a Brazilian man, Thomaz Morias, who was struck by lightning and can do even more of these miracles. But the objective of this book is to find a common principle, not collect more data.

Scientists in the former Soviet bloc assumed these effects are caused by a natural bio-energy which people project outward on objects. It is unlikely that energy projected from the outside causes the movement, but rather that the idea do inspire the object to move is innate. It is thought projection, which is clearer in humans, which causes the materials to react from the emotional energy existing in the materials themselves. In the case of Thomaz Morias,[3] he has only to think about objects like spoons bending and they bend. It is not bio-energy—it depends on the proper emotional frequency of the sender to affect material objects. This explains why only a few special people can influence materials. People with this special emotional capability have the gift that is usually unpredictable and a passing stage in their life. Recently a Chinese psychic Zang Baosheng became known for similar powers.[4]

Skeptics investigated all this critically, looking for fraud found some cases of imposters. But a few bad apples do not prove that all apples are bad. And if the skeptics could not disprove phenomena, they at least sneered at it to show their sincere disbelief and their loyalty to materialism, thus displaying a narrow mind.

Another confirmation of the innate ability of the human mind to organize is a contribution from "gestalt psychology." Experiments with perception show that the human mind inadvertently organizes all impressions into meaningful gestalt. Organization is a higher level of consciousness, not just of the human mind. Amorphous energy without innate intention to organize, would be insufficient to explain the complex organization of the world all around us. The design and activity of beings express ideas what they mean, they are ideas in materialized gestalt, that reflect their purpose in the scheme of things.

Since we already know that the biological energy on the molecular level is electrical in essence, this means that organization by ideas is basically an electrical wave phenomenon, indirectly implying that electrical waves are emotionally influenced to organize. Why are physicists so impressed by the colossal energy that pervades the universe of matter that they could not see a connection to human emotions? Is it because physical experiments confirm the mechanical nature of energy? Or is it because their experiments are set up in a mechanical way to which they are committed by an oath of loyalty? In fact, if we humans are facing challenges we can also generate emotional energy to meet emergencies. One could say there is no comparison between our insignificant emotional energy and nuclear energy. It is like comparing a raindrop with the ocean, but in principle it is the same water. Since physicists believe only in accidental occurrences, motives are meaningless ideas and misleading at best. Directly perceived, that may sometimes appear true. Ancient people believed that electrical storms with thunder and lightning were emotional expressions of angry gods. This was a misconception. Nature is impersonal, yet it can still be emotional energy. It would be on a different mode, the correction of electrical imbalance, which in human terms would also cause conflict and anger.

If ideas can generate energy and organize things and also cause thoughts and images appear in our minds. That means ideas can transcend the separations between things and beings. We transfer ideas to others, which is what we normally do when we talk to each other and share ideas, face-to-face.

If the universe is an emotional process then physics would be wrong to assume that it is a mechanical system. From calculations made by physicists recently, our sun should have used up its energy long ago—unless it somehow can create more energy. This must have occurred because it is still shining and science is frequently extending the life of our planet by millions of years. But what could have inspired the sun to create more energy? It could be from the feeling of being needed and useful, even happy, to keep shining on its planets, including the blue planet we call Earth. We know from psychology that human beings have emotional ups and downs, and can generate positive emotional energy when happy and negative energy when unhappy. Unhappiness is disorganizing and effects health and life span. What we see with improved telescopes happening out there in the physical universe is similar in principle to what happens among living beings, namely disorganization and reorganization, death and birth. There is no indication of a grand "one time singularity," as a big bang explosion of the entire universe, but continuous change where only individual stars explode in super novas, and eventually become reorganized in new galaxies.

One could ask what have humans and atoms in common with stars and galaxies? They all are organized by ideas of meaning.

Now let us think about the mysterious phenomenon of psychokinesis, a field under investigation by parapsychology for the last half-century. Parapsychology has discovered that a special state of "emotional excitement" is essential for psychokinesis to happen. This opinion is the result of the investigation of many, otherwise unexplainable events which confirm that emotions in a heightened state of excitation can transfer the excitement to material objects and cause a reaction. In the case of psychokinesis, the investigators arrived at a similar conclusion after studying many emotionally "distressed" persons who cause unusual responses in material things. I do not think that psychokinesis happens only in a few frustrated individuals. It occurs frequently, but we try to come up with other explanations for these happenings to remain physically real.

In a few recorded cases, the accumulated frustration of unhappy individuals has caused ceramic objects to fall and break, rocks to fly, or telephones to ring, etc. This seems to often happen unconsciously to a frustrated individual who never did anything destructive directly. To this may belong the witch-hunts when persons were accused of bewitching with evil intent, and the rare cases of spontaneous combustion during which frustrated humans burn from the inside, causing their own death.

Obviously in extreme emotional states, transcendence can occur. Psychologists must sometimes acknowledge this occurrence when time is transcended, telling persons what they are not necessarily out of their mind, that they *can* sometimes predict events before they happen, that they have a psychic gift. If they can keep their composure it may be a case of transcendence. It is also a fact that the mentally ill, when they begin to improve, frequently are predicting their own cures in dreams, as they make a turnabout to a positive direction. Rare as psychokinesis may be, to ignore it means to ignore an important clue to the deeper understanding of the principle of reality, which in normal life experiences means little to average people who just want to live unperturbed by complex phenomena and thoughts.

Among the strongest defenders of the material world concept is the medical profession. Whenever incredible stories of successful "psychic surgery" spread to the public, they want to check it out and unmask it as fraud. With this motivation a group of American doctors went to investigate what they expected to be cases of clever deception.

What I am referring to is the report by a group of medical doctors who went to Brazil sometime during the seventies to unmask the already established "fame" of a simple farmer. The psychic surgeon whom they met in person was an uneducated, friendly individual who earned a small extra income in the way of gifts from grateful persons on whom he performed "surgery." He mainly made his living from working a small farm. Charging a fee to sick persons was unthinkable for him. He was known to be deeply religious, holding a mixture of Christian and pagan beliefs common in Brazil, partly with roots in Africa.

The surgical demonstrations he gave for the visiting doctors were performed on local patients with only his hands. Since his arms were bare, it was clear he could hide nothing up his sleeves. He also diagnosed and removed a tumor from the neck of one of the American doctors who had come to investigate him. What they saw was magic to them, strictly a mystical performance that belonged to another time or another world. Incidently, they filmed everything to make sure it was not an illusion—that it was real. This confirms that physical reality can be transcended with a particular emotional state of selflessness which is also emotionally based, but a positive, confident form of transcendence which is also necessary in fire-walking. The news media reported the findings on the psychic surgery done in Brazil. The doctors did not try to deny it. Public television showed a program about this phenomenon which I happened to see. In our time, people are used to hearing all sorts of things which they can believe or not. For the average over-informed person, this report was taken as just another confirmation that nobody knows what nature is, in a world that has been recently called "chaos." Besides, we are so used to stories about faith healers that we no longer think further than the medical implications involved. The average person usually misses the bigger philosophical consequences behind it, namely what transcendence means to the current material concept of reality.

The fact that faith healers occasionally are successful is recognized, but for proof we need healers with "consistent results," and they are hard to find. The practical answer is a teachable technology, such as a medical profession. For all practical purposes in our biological age, the medical profession is specialized for unusual demands and immediately accessible. Whether one uses a knife to operate or only hands is not enough reason to travel to remote areas of Brazil or the Philippines to have a tumor removed.

A well-known psychic healer was Edgar Cayce of Virginia Beach, Virginia. His method was to put himself into a deep trance and then have others ask him to diagnose various clients, who could be living anywhere. Their address was all he needed. Without personally meeting them, he would diagnose and prescribe a cure, and he helped many. One should also mention religious faith healers, like Katharine Kuhlman for example. She admitted in an interview that the religious beliefs of the persons she was able to help during her career had no bearing on results. Among her successes were also persons without religion. Apparently healing depends on personal rapport, she said.

The laws of physics and chemistry certainly are not rendered invalid by occasional cases of transcendence. Medical technology is useful since the human body is composed of chemicals. If physical treatment fails, the next step is the psychological approach, an already normal procedure. Our thinking has changed since the days when sorcerers drove evil spirits out of the body. In general, spiritual causes were assumed and this continued until the eighteenth century when witches were occasionally still burned.

Now with the acceptance of psychokinesis we are back to spiritual explanations, but our understanding has changed. We gave up burning people in a cleansing fire. Psychology, in particular "parapsychology," has arrived at other solutions, namely helping frustrated individuals to calm themselves, mainly by changing their situations.

Let us take the well-known case of a teen-age girl who lived in Rosenheim, Bavaria, where she was employed by a lawyer's office. For unexplained causes, a staggering number of telephone calls to the "standard time" number appeared on their bill, as many as three calls per minute, all requesting the exact time. The telephone company sent their electronic experts to search for technical causes, but everything seemed in order. Everyone who worked in that office denied having anything to do with the calls. Besides, it was physically impossible to make three calls in one minute on a single phone.

The newspapers printed the story as a mysterious occurrence. This attracted the attention of a parapsychologist who speculated that it could be a case of psychokinesis. When the new girl was asked to take a day off, the barrage of calls to the standard time number stopped. When she came back the office staff watched her carefully. Although she never made a telephone call, the calls started again. This solved the mystery. Obviously she was frustrated in her office job and wanted it to be five o'clock so she could get out of what to her was a frustrating confinement. This case was solved by elimination. The girl was let go—no doubt she was glad—and now is a happy mother with a family of her own. We understand that frustration can cause the mind to act in magical ways and effect mechanical and electrical devices. Many mysterious events happen which resolve themselves and never reach public attention.

Another well-publicized case of psychokinesis in our own time occurred when a newly built home in the outskirts of Tucson, Arizona, was occasionally pelted by a hail of stones, breaking windows and such. The police were called and they set up surveillance. The surrounding area was semi-desert with many stones on the ground and sparse bushes, with no chance for stone-throwers to hide. A parapsychologist finally solved the mystery by elimination. The family had a teen-age son who was frustrated having to live in this new house away from his friends. The boy was never considered the culprit since he was always inside the house when the stones came flying. After he was sent to live somewhere else, the bombardment stopped.

In both these cases a frustrated teen-ager was the cause. Whether puberty is part of the emotional frustration is open to speculation. But who threw the stones when the son was inside the house? Or is it conceivable that a spiritual part of him did it without the other part knowing?

Out-of-body experiences are known to be possible although scientific proof cannot be provided. Yet the results of moving an adolescent to another more congenial place solved the problem. This means we arrived at trial and error, by finding out what is real by elimination. In order to arrive at a common principle of nature, elimination seems the best method. The first we have to eliminate is the mechanical part of physics. It is an artificial technology and therefore a sideline that leads to a dead end. Quantum wave theory has separated from the material concept of nature and indirectly confirms that the essence of reality is a common emotionality.

Emotionality is the common essence that runs through all levels and manifestations of reality from conscious humans to inorganic materials to invisible electronic wave energy, which is emotional excitement in its purest form, namely light. On the human level organized emotional energy shows full self consciousness. The human mind and life experience is one of the higher expressions of the evolution of emotional consciousness. The human being is still basically emotional but can become reasonable if it can control its emotions. This is the stage on which we humans presently exist. We try to control our emotions to become more rational, within limits, than is possible as emotional beings in an emotional universe. One can conclude that the ultimate essence of reality is emotional energy that is partly conscious and has a tendency to organize itself into a multiplicity of self-expressions. On our level, the most striking feature is organization into meaningful ideas manifested as physical gestalt, in what essentially is a wave phenomenon suggesting solidness.

2

The Principle of Reality

THE FIRST CHAPTER BEGAN WITH A CRITICISM of the present attempts by physicists to unify all reality into a single "theory of everything" as a mathematical equation. The basic concept of this mechanical view is that everything in the universe, from atoms and molecules to galaxies and stars, is held together by gravity. However, nobody could ever confirm that gravity is physically real, only that it is an idea to explain an effect.

Around 1925 another theory called Quantum Wave Mechanics was developed. It is based on the discovery by Max Planck that energy occurs in bundles since a particular repetition of multiples of a constant number always recurred, indicating that somewhere a basic entity must exist which he called the quantum. The level we are thinking about is invisible energy. The quantum is only confirmed in groups of waves that appear as bundles or particles. But these wave bundles, compared to known things and beings on other levels appear pre-organized for further organization. This deduction is based on existence of organization on higher evolved levels—although with invisible energy, this is an assumption based on the existence of a common energy. Although organization is not the basic principle, yet there is something which inspires organization and causes disorganization. Both are based on emotions. Incidently, further splitting of subatomic particles apparently only causes the waves to assume a different form of reorganization. I think this shows the unbending intent to reorganize into something, because at this stage the intention to evolve into something of meaning is already there as a compulsion. It is the recurring intent to organize waves that express themselves as meaningful ideas on clearer levels of reality, where motives are discernible. Such ideas evolve into meaningful beings and seem to occur on separate levels in different modes, beginning with the electronic level of wave frequencies, to the level of atoms, to molecules, to crystals and to cells. From cells, larger organizations evolve into complex beings with conscious orientation, including self-conscious humans. One could also understand it as an evolution from a low level of semiconsciousness to full self-consciousness.

At this point many would object because I assume that the material level is already conscious. This assumption I have based on the reaction of material substances caused by emotionally frustrated individuals, known as psychokinesis. This reaction implies that quanta are emotional in essence, that emotions cause the waves which ideas organize into anything and everything. This again is confirmed by the discovery of emotions in pure energy by experimental methods. Experiments alone are not enough, but with all the other levels also showing emotionality—that is convincing.

Organization is also apparent in the macro-cosmos in entities that are separated by vast distances, like stars that have a flock of planets around them that loyally travel around their mother star in elliptical orbits defying the concept of mechanical gravitation. And some travel in an opposite direction, with opposite spin, and at distances that exclude gravitation like Uranus, Neptune and Pluto.

In general, from all these patterns a purpose is implied. One could make a comparison with organic parallels where mothers have a flock of offspring following them around, and when fish return to their place of birth and pigeons return home but without a fixed orbit, after a random journey in the world they return to their place of birth. And if one were to say that humans do not return home, most still feel some loyalty to their ethnical roots. But if many disregard their origin that confirms that somewhere estrangement occurs and one changes. That is how evolution makes sense—one separates and became estranged and this is also reality. It is the struggle between loyalty and independent living. Whether conscious or unconscious, there is a sense of belonging in nature. Among humans the conscious organization continues beyond families into larger groups that are held together by a common interest and by ideas which give meaning to their lives.

In essence, on our level ideas bring us together and determine what we do. In principle, that may be what happens everywhere, and that includes the stars and planets. This means that loyalty is an emotional idea which cannot be recorded by instrument, because our thoughts cannot be measured. Yet our thoughts not only exist in thought but guide us through life. Again, the most important determinants are nonphysical and often compulsive.

Organization always starts with indivisible entities—on the human level with the individual. As natural individuals we also cannot be joined together physically, but will organize because there is an advantage in helping each other and we are attracted to be together. In this way families and states are natural organizations of individuals, yet remain separate individuals and can disorganize if they dislike each other. Let us compare this with atoms. Atoms for the longest time were considered indivisible until physicists succeeded to split atoms which released energy. But to join atoms together by fusion did not work. For fifty years they have tried it and have raised temperatures to that of the sun but still failed to achieve fusion. Therefore scientists concluded that all the complex atoms that exist in the universe—that means everything above what the sun can fuse, were created with much greater heat, namely in a singular cosmic explosion and therefore they presumed that atomic creation by fusion took place in the first second of the big bang.

This a typical mechanistic conclusion based on the secondary organization of atoms on the molecular level that does not apply to the subatomic level. In nature, organization is meaningful and hydrogen and oxygen atoms have organized to form water, a substance on which organic life depends. But it leaves the nucleus of atoms separated—only the electrons connect—similar to secondary organization of human individuals in a cooperation of groups. The reason for human organization, we know, is in the advantage of a larger organization. We cannot know what atoms feel or think, but if nature is meaningful then meaning is a motive for the secondary organization of atoms into larger molecules and all other organizations which allow individual atoms to exist as separate entities that make up our physical world in which only the electrons connect.

Humans learned from experience that fire and heat changes substances and developed chemistry, but were misled in thinking that atoms also respond to heat. Heat and fire do not destroy atoms; they are organized by ideas of meaning. Because chemistry succeeded in creating some artificial compounds that in nature do not exist, that still have use only for humans and otherwise result in pollution, physicists are trying to apply heat to fuse atoms as if they were molecular compounds. They seem to think that the entire universe is a chemical laboratory on the material level.

From the human level we know that organization is influenced by ideas of meaning, a mental process. Physicists seem to ignore this possibility. And they also ignore the possibility that ideas are expressed in all things and beings that exist in nature, a concept that was recognized by Socrates and Plato two thousand years ago. I therefore think that atoms, as basic building blocks of the physical world, are materialization of consciousness into meaning. Below the atom only immaterial wave bundles are pre-organized into frequencies of energy which exist on an emotional level with basic ideas in fields of consciousness. Or in physical terms as quanta fields organized by ideas into simple meaning which evolves into complex ideas, but slowly. To break up atoms is easier than to organize atoms. Just as it takes only a second to shoot and kill a human being, it is harder to create one. To organize and nurture one human takes many years, especially with the idea of becoming a reasonable human.

Obviously physicists and experimental scientists have given up on reason. Instead they are concentrating on discovering the existence of mechanism of nature and with that, learning how to build mechanical contraptions that are useful to mankind. These may be useful to people in the short run, but in the long run become quickly obsolete and, as refuse, pollute the environment. One can see that this is artificial and the wrong direction and many intelligent humans already know this. I am neither the first nor last to say this and it has not impressed the mechanists in their single-minded intent to discover more ways to exploit nature to make more money for themselves in lifetimes that are too short to ever spend all their wealth.

How can one person assume to tell others what to do? I have no illusion. Exploitation will continue, but as one person, I can only suggest an alternative. That is why criticism alone is not enough. Unless a better explanation is given nothing will change. This fact was on my mind for a long time. I did believe that ideas organize everything because I had read books on philosophy. But philosophy was demoted to a handmaid of science. It had lost credibility because ideas directly projected on objects create nothing, but indirectly they might. I wanted to make a convincing argument on how organization creates things through ideas. I was thinking along these lines when an idea occurred to me that convinced me the principle of organization was understandable, but indirectly as vision, because physics was hopelessly lost.

Let me begin at the beginning, because it took a whole lifetime to come together in my mind. Besides, anecdotal thinking is more natural and interesting than abstract formulas.

I have sometimes had revealing dreams. I was then just beginning to take an interest in the psychology of the unconscious which in dreams see a connection to the conscious mind which gave me a different perspective of reality. I had an interesting dream which I would like to mention. I found myself in a garden, similar to ancient Greece. The people were dressed in white shrouds like in Greece or in another world. I knew that I did not belong there and stayed in the background. Besides, the people were paying attention to a procession that took place and not to me. Some were carrying a person on a stretcher. It was a woman who sat up, and the crowd was applauding. I looked around and noticed a stone statue, such as one would have found in ancient Greece or in a cemetery. The statue represented a beautiful woman whom I found very attractive and graceful. As I looked at her face and into her eyes, suddenly one eye became magnified and revealed the actual substance, which was stone, showing the particles of the grain as its true essence.

With this vision I then awoke, speculating about the meaning of this unusually clear dream. The first part, the procession and the woman on a stretcher, I was able to explain in the next few days. A friend sent me a photograph of his mother's funeral. She had died after a long battle with cancer. But the other part, about the eye of the statue, had another meaning. Obviously the message was that the external image is not the same as the true essence. I had admired a human image of beauty, but indirectly was opening a channel to the material consciousness by admiring the image. It seemed the dream wanted to remind me that materials also have feelings.

I later read about what others have to say about this type of dream. One dream book said that if a statue comes to life it means an idea will soon come into my awareness. I learned this about the time that my interest on the subject of organization started, particularly on the cause of the organization of cells in the human organism.

Later it became clear to me that the first part of the dream should tell me that I was on a spiritual level. Therefore, the message from the statue was of transcendental significance which I took as confirmation that materials have emotional feelings and are remotely connected to us. But it also reminded me that external form is not identical with the consciousness inside.

I already knew that the physical impression is insufficient to understand the essence of anything, but it became further enhanced through this dream. It also explains why any philosophy based only on external appearances and physical measurements is missing the important part, the essence of its meaning. This is precisely why materialism fails to find meaning in the universe because it ignores consciousness as the essence of reality.

In our time when the mass media is influencing the thinking of people, the words of Marshall McLuhan that "the medium is the message" has some truth. Without sufficient life experience, younger persons take the messages of the media for real. This is particularly true for the emotionally charged shows disguised as entertainment, suggesting that uncontrolled violence is normal. In fact, uncontrolled emotions are primitive, on the level of animals, which, of course, encourages our latent criminal tendencies to act freely. In our materialistic civilization, the survival of the fittest is accepted, encouraging exploitation in business and commerce as the natural way. That may be true for nature and be wisdom for animals, but whether humans should take this message as guideline for life degrades our intelligence. For us it is different: As we treat others so will they treat us. A friendly and open approach will result in an equally friendly relationship. But this only makes it easier for cunning people to deceive someone who does not think for himself and is not aware that a person's exterior can be misleading.

If sensible thinking is the basic requirement and life experience (which I have) helps, there is no reason why I should not be able to write a book with a message. Whether my ideas are correct will have to be confirmed by reality. Having been blessed to live in interesting times, such as the Second World War and what has followed since, has expanded my awareness. Allow me to go back into my life to explain why I came to think that there is proof that a common principle of organization is evident in nature. This can best be understood when I mention the personal experiences by which I arrived at this belief.

In 1943, when I was eighteen years old, I was living in Central Europe in the northern part of Bohemia, in a region that had been annexed by Germany in 1938. I was drafted into the German army and matured quickly after surviving eighteen months of the senseless carnage called war. During this time I lost my belief in the intelligence of mankind, knowing that it is difficult to control the emotions of the masses once they are fired up. I decided that I would never become a follower and let others think for me.

I was turning twenty when the war ended in unconditional surrender and I became a prisoner of war. My parents were expelled from their home and their little farm in Bohemia and driven to the German border, with only the clothes they happened to be wearing. This was still a better fate than the degradations many others had to endure by hateful people who only asked what language you spoke. Rape, torture and murder were common for the losers to suffer before some measure of safety and order was reestablished.

I found my parents and we lived in primitive conditions near the destroyed city of Frankfurt on the Main River, which slowly dug itself out of the rubble. After the School of Engineering reopened, I applied as a student, but I was no longer an attentive listener. My mind often wandered in day dreaming. The seemingly endless calculations on mechanical trivia were very boring. During the second semester I took the advice I heard in a speech from a professor of mechanics. I remember he said that not all persons are cut out for this profession. Once a student's mother asked him why he had failed her son who already had two patents to his name. He told her that if the son had an intuitive mind, he should be an inventor. As an engineer, rigorous attention to details is required. Since engineering is based on accurate calculations, a mathematical inclination has to become instilled as a habit. Engineering is a profession that stands for exactness and dependability, but not necessarily vision.

That was my last day studying mechanical engineering. It took half a year to process my application for immigration and by the first of June 1951, I was on a ship to Canada to seek my fortune in a more interesting life experience. The prospect of wasting my time calculating trivial things all my living days had scared me. I had given up hope that Europe would ever become united because the ethnical differences were deep. And having been expelled from what became Czechoslovakia just for language, unity seemed even more hopeless. At least in North America, where the people have outgrown their ethnical differences and all learn a common language to understand each other, cultural differences actually become an enrichment of the cultural life. I am used to thinking for myself and found the freedom from regimentation a welcome relief.

Because of my mechanical background, I found employment in machine shops, working in toolmaking and precision inspection. One has to earn a living somehow. Penniless, I came with the advantage of already speaking English. At least my work left me free to think. I followed many interests. Reading books on philosophy and psychology was among my entertainment. As the saying goes, man lives not by bread alone. We all live in different fantasies. Certainly I live in my fantasy, speculating about reality in the indirect company of ideas.

Throughout the long history of philosophy a large number of humans have always believed that ideas are the essence of consciousness and inspire the dynamics that keep the world in motion, or emotion. The first two years that I lived in Toronto I was single. Then I found an interesting companion, a versatile woman who had also been expelled from a region of Germany that had become a part of Poland. Together we raised a family and kept ourselves away from political attachments.

Whether anything qualifies me to write a book about the psyche of the universe is open to interpretation. Belief is not the issue; to think for oneself is the requirement for reading this book. What I intend to say is not written in stone as we are only human.

Just because I started this book with current theories in physics, like the wave theory, my own thinking at that time was far from physics. I read a lot on the subject because I needed to prove that mechanical thinking is unnatural, that emotions can effect the organization of wave patterns and that transcendence is a logical possibility. Otherwise, physics did not interest me. Actually, thirty-five years ago when my interest in organization began, I was uninformed about physical theories. In fact, most were yet unpublished and restricted.

It should be mentioned that at the time I came to Canada, I was a single man of twenty-six. Most of the knowledge I now have, I acquired over the forty years I have lived here. I had the idea about a common principle after I lived here for almost ten years. It came to my attention that physicists were claiming to be close to discovering the principle of the universe and would soon have a mathematical formula that would explain reality—a theory of everything.

With only a general background of philosophical idealism, this boast seemed an arrogant imposition. I believed if there is a single principle it is found in the "secret" of how organization works, which involves conscious orientation but never is a mechanical concept.

Sometimes I thought about this principle of organization as a novel idea, yet many others must also have observed it but considered it a mechanism or secondary, or whatever. In fact, many biologists already studied the subject of organization of cells in microbiology as a chemical process. Eventually someone would explain why cells organize into meaningful life forms from ideas. But then I also knew that with the mechanical concept which dominated biology this would not happen. The idea would be ignored by the dominating mechanists as old stuff, already disproved as spiritualism in nature which progressive scientists have relegated to the past.

Proving that ideas organize everything is very difficult because major evolutionary changes happened far apart in time and visible transformations did not occur in our time, and experiments that were conducted were mechanical; nature never evolved like it happened in laboratories. Scientists may think that they are near the truth, but needless to say, the theory that thought could influence organization and growth had failed in experiments. They were all convinced it was a waste of time and a useless pursuit that showed no results because ideas create nothing. They had proven this to themselves. Because nature does not respond to our mental influence and wishes, that ends the dispute.

I had read books on psychology dealing with the unconscious and knew from psychokinesis that mind or emotions connect all things and beings, I expected that a breakthrough might eventually come from psychology, certainly not from a mathematical formula, like the Grand Unified Theory of Energy or the Big Bang theory of gravity. The only energy that can come near to explaining meaningful organization is emotional energy and that field belongs to psychology. Psychology is going to solve it eventually, but since I was working in tooling on automation in the production of automobiles, what could I contribute? From the information that I had, I should have been a convinced mechanists. That is where I was in reality—just an average guy who liked to read interesting books. My wife believed that philosophy was a hobby to counterbalance what I did for a living. I resigned myself to reading about such fields and to enjoy living as much as possible in my situation as a family man, which was fair enough.

One evening a rare coincidence occurred. A public demonstration of the nature and range of hypnosis was given. I was able to attend because I was on day shift. My knowledge of hypnosis was limited, formed by standard opinions typical of our time. As I saw the various effects that took place in the trance state of hypnosis, I noticed that the unconscious allowed various suggestions to influence the organism of a normal person. This meant that the natural defense, which normally protects the organism from unwanted influences, was bypassed. Although the influence of the trance was limited, it showed that influence on the organization of the cells in the body is possible. I had been thinking that an explanation of organization of the human body would reveal the underlying principle of organization, and now by a rare coincidence had found an answer. Suggestions during trance are reaching the unconscious level where the organization of the body is situated, and suggestion actually can have a temporary influence on the millions of cells in the human body as the cells follow the commands of the hypnotist. It was to me as if the secret of the organization of nature was revealed, which to the others seemed a questionable curiosity soon to be ignored or considered entertainment.

Medicine also assumes that the human body is an organization of cells. The idea of a "cell state" is used as a metaphor to explain functions and responses, but they believe that the body is a rigid regime that resists influence. Although the visible influence of actual organization of cells during hypnosis was minor, consisting of no more than raised blisters on the skin, it meant that cells did respond to a command to reorganize themselves to form a lump. I had found the needle in a haystack and was excited.

The rest of the effects of hypnosis impressed me as well, namely the temporal control of the muscles in the body. It was amazing that this involved millions of cells that could be influenced to contract in unison. Next, the effects of hypnosis on the imagination, particularly the ability to cause hallucinations in the mind to be experienced as real, convinced me that suggestion is the key to organization. And what is being suggested are ideas.

It was soon after this demonstration of hypnosis that several connections occurred to me. I reasoned that if on command blisters could instantly be raised on the skin, it meant that in principle hypnotic suggestion can influence the organization of cells in all living beings. Because influence is only possible in trance through suggestions to relax and to feel tired, hypnosis is a sneak entry into the unconscious organization of a cell state. This means, in trance or during sleep, the unconscious is unguarded and open to suggestions. The fact that suggestion, after slipping by the critical mind, can influence the organization of the cell state, is already of philosophical importance. Just because we humans are limited in our influence and can cause only minor effects, this only means nature does not trust outside interference. It does not mean that suggestion is meaningless, only that we are restricted.

Because human beings apparently have more than one level of consciousness, the influence is restricted to the upper levels within the reach of verbal understanding. There is much more beyond the reach of language that may effect organization—emotion for example. The mind also has many doors it can close to guard itself from unwanted suggestions. This means the details of influence on organization are unknown but the principle is known.

The medical use of hypnosis, although interesting, is of less concern to me because the issue is the explanation of the principle of reality, and clarity is better served to keep unnecessary details for later. The unconscious is resisting interference and normally expels all foreign suggestions after the trance has worn off (this also happens in organ transplants, which can be implanted but later are rejected). There are many parallels pointing to the protection of organization on each level. On the organic level it is proven. On the mental level it is experienced as hallucination and beliefs, and on the material level, it is emotionally affected. If there is only one principle, one could use the proven effects to explain the unprovable. At least that is my opinion, but why not give skeptics a chance to object? I could be wrong about details.

Hypnotism is certainly limited because people usually awake from trance if ethical barriers are reached. Of course, if one has no ethical barriers, such as criminals lacking any moral conscience, one may not resist unethical suggestions. Yet not only criminals accept unethical commands, but also other humans who from childhood are desensitized by exposure to seeing people killed every day on television and therefore think such behavior is normal.

Apparently during the last fifty years of experimenting with hypnosis, we have never gotten much further than what was learned in the first fifty years since its discovery. We know, therefore, that we have only a limited influence, and that for only a short duration because the unconscious identity of a person restores itself after awakening from the hypnotic trance. We also know that the design by which the body is organized is partly inherited from individual ancestors and partly acquired from experiences in life. In any case, nature limits our influence on the organization of its beings because suggestion works only indirectly via the sub- and the unconscious where many different barriers protect the inner organization from willful outside interference. This is a wise provision of nature—to avoid confusion and to keep the countless beings that exist in the world from confusing each others' identities.

Thirty-five years ago I was ignorant of most details and subsequently read all I could find on hypnotic suggestion. In the demonstration of hypnosis, I had to assume that during the trance the muscles of certain parts of the body contract selectively, except for the heart. The heart, also a muscle was unaffected, as was breathing. For all cells to lock would mean certain death. This is a further indication that the body is very selective. Even in a trance the body knows what cannot be allowed.

Checking my thoughts again, suggestions must influence millions of cells in the body that temporarily obey commands given from outside—during trance, that is. I knew then that suggestion is the key to the secret of the organization in all of nature, and that multiple separations, external and internal, are protected levels against interference in their separate existence. This means all separate entities are part of this provision. Organic or inorganic, they exist because of separation of meaning and identity.

But I had to ask myself, why did all the others who played around with hypnosis for a century not notice the far-reaching implications of suggestibility on organization, and long before I did, since I am only a dilettante? Of course, indirectly I learned from the discoveries of specialists because they have published their findings. I just do not have a degree to display for my credentials. I would not have agreed with the mechanical concept at the basis of my feelings so that would not have happened anyway. From what I could read in books written by professionals, apparently organization and separation did not mean as much to them. In microbiology where scientists analyze the human genes and the molecular structure of cells, they talked about mechanisms, not selective responses to the suggestion to become different. Why did these accredited professionals ignore suggestibility? Was it because they imagined there is a fixed blueprint in the genes and that everything is mechanical, and therefore suggestion is ruled out and the presence of unconscious intelligence must be categorically rejected?

Obviously, suggestibility is present in all levels of nature and is necessary to explain the orientation of organization to meaning. I was excited about the idea of "innate suggestibility" in all things and substances. It was like a revelation and gave credit to a general background in the history of philosophy, which when I was a teenager had opened my mind to many possibilities. Far from detailed knowledge of philosophy, all I had was a general overview. I knew that philosophers from Kant to Hegel and Schopenhauer believed that ideas organize and identify everything in the world. This is now verified. However, verification was not as they had expected—not by innate will to assert or by the dialectical principle, even less the will to power of Nietzsche, but by innate suggestibility to meaning by selective intelligence and reason. This is expressed in the intelligent organization and the direction of evolution from impulsive emotions to self-understanding and compassion for other beings. This may sound like an optimistic conclusion which a skeptical thinker would hesitate to make.

This understanding of the influence of innate suggestibility on organization felt to me as if an idea was planted in prepared ground and convinced me that something important with a vast range of implications had come to my understanding. Perhaps the dream I had was preparing me for this insight. I was not greatly bothered with details and held the opinion that reason is the ultimate criterion of what is real. If the main idea makes sense, the details will take care of themselves. This belief I never gave up, even in mechanical systems reason applies.

I do not agree with those philosophers who hold the opinion that men make up their own philosophy. Nature exists, is the ultimate system which we already know has evolved from one common substance or essence, which I, like most are trying to understand. That is the issue in our time: understanding the principle of reality. That a common principle exists must be assumed and eventually understood.

Let me expand on other feats of the demonstration of hypnotism that I saw. I observed how a man in trance, totally rigid, stretched out and lying on his back, could remain suspended between two chairs, the back of his head on one, and the heels of his feet on the other, without slumping down in the middle. This is already beyond normal strength. In addition another man was sitting on his middle as an example of the extent of the influence of suggestions on the cells of the body. We know that in a normal state of consciousness such locking of muscles is impossible, but in trance this suggestion is accepted and implemented by the body. The explanation is that in a normal state of consciousness, our reason can not believe we have such strength and does not risk it.

In normal consciousness the cells will not obey such orders. Only when the conscious reason is bypassed are unreasonable commands obeyed. We are normally skeptical about incredible ideas that could cause us injury, the unconscious apparently is setting a limit and ending the trance when it becomes life-threatening. Protectiveness can explain why in the conscious state suggestion does not work. Our reason has ideas of what is possible and usually is careful about accepting new suggestions. This also can explain why people, even if they willingly agree to hypnosis, can still subconsciously resist trance. It can explain why it is difficult for certain persons to achieve any form of trance, let alone a deeper level, if they fear surrendering control of their life to another person. That explains why hypnosis is unpopular, because we make it difficult.

If it were easy to suggest whatever one wishes, then the world would be impossible to live in. A system based on suggestion cannot exist without protection against all other unwanted interferences. This means a world created by suggestion would have to be separated and divided into countless levels and distinct individuals who are closed within themselves to outside interference, at least in their internal functions and organization.

Still for further protection, they would be separated by major changes of levels in the physical structure and by entirely different modes of being. These modes would be as different as the organic from the chemical level, and the electrical from the nuclear mode, all indications that separation of identity is fundamental in nature. In fact, separation is the actual pattern on the electronic mode, besides the separation of waves, ranging from micro- to macro-wavelength, and mind frequencies as well.

Presently we can actually see with microscopes how cells divide and organize into living beings, beings that have meaning for themselves and that defend the identity into which they were organized. Nature, in fact, already follows the intent to differentiate itself from the start, which led to the estrangement of entities that formerly were related, and eventually finds them incompatible, even intolerable. This estrangement continues by further separation and differentiation, which frequently lead to collisions of emotions, to violence and to organized war. Wars among humans are usually brother wars. Yet I am aware that all this is a generalization about an obviously very complex reality that at first glance appears chaotic. It is a deliberate simplification in order to bring strange ideas under one denomination, namely innate suggestibility that has evolved to self-knowledge.

As I have already mentioned several times, it seemed logical that with the progressing differentiation that drives everything apart, something must be presumed to hold the universe together. This should be the unconscious, without our direct knowledge, because we are meant to believe ourselves to be different. Unconsciously and indirectly we know that we are of one essence, because in dreams we are all interconnected but still selectively and unconsciously sort meanings in different ways. We are indirectly aware that we are not completely autonomous, but are connected emotionally to the whole by a universal consciousness out of which we have evolved.

This same awareness through which ancient philosophers called the world the "many in the one" is a system of nature that is separated and interconnected at the same time. Dreams and extrasensory phenomena demand such an interconnection as explanation. Channels that exist must allow the feeling of free choice to experience life separately—this could only mean indirect channels—in order to leave us the belief of autonomy, to allow us to think we have control over our lives. Twenty-five hundred years ago the philosopher Heraclitus of Ephesus is reported to have said, "We are and yet we are not," which in such a context makes sense, but otherwise could be called illogical.

Thirty years ago, I could think of such a suggestible universe, but I was still uncertain about many details and concerned about the possible consequences it could have on people. Therefore, I kept it to myself until I could get it written right and some of the details more clearly defined. But because of my job working on machines, it might never be possible to put all this down in readable form. Against the dominant materialism that rules the world, it seemed a tall task for one man in the wrong place. But I quickly ignored this after a few minutes.

If the world is created by innate suggestibility from emotions which transformed into waves that manifest as solid at certain speeds to give the impression of the material world, this makes sense to me, but to convince anyone else is a different subject. I had never written a book, and who is going to listen to an unknown person? People might politely listen perhaps to a nuclear physicist, but not for very long. Even physicists might have a problem convincing average persons of strange ideas. Besides, the majority of scientists still think that suggestion is a completely unimportant side effect of an accidentally evolved, meaningless nature in a material universe. It is a question of credibility, and I am the last person to have credibility.

It took me years to put all my personal concerns aside. Who needs the aggravation from fanatical persons who might think I want to destroy their beliefs in their worlds? In fact, since many worlds are conceivable in a universe of suggestion, all is possible as long as it has meaning for them. It supports multiplicity of belief systems. This philosophy is positive, and is a hopeful perspective of a meaningful and interesting universe. But I still had doubts about why all the others who actually use hypnosis in their professions never saw the importance of suggestion. Were they all fixed by a belief in the laws of nature already cast in stone by science? Is therefore suggestibility only a minor side effect of physical reality because they all agree it is useless? Is suggestibility therefore a contradiction of their reality where all things are explained by mathematical truth, where consciousness is irrelevant, except when it comes to their own conscious minds? Is the most important question they must make an exception for themselves and to add the entire organic level as freak accident to their material universe?

I cannot believe it, but then, thirty years ago, in the back of my own mind there was still a lingering doubt about suggestibility itself. A critical point was unexplained that could bring this theory down like a house of cards. It concerned the fact that hypnotism and suggestion could also be used quite frivolously for public entertainment. Practically anything could be suggested: Total nonsense, insults and degrading acts, including murder. No serious philosopher would touch suggestibility. It lends itself to everything and is perhaps nothing, only a technique that is empty of content. This means it is the content that is important. This reminded me of mathematics, which can be used for anything, but depends upon whether one calculates things real or imaginary. It is, as Bertrand Russell said of mathematics, a tautology, and so is hypnosis. The issue is the content and whether it has meaning and is real. But what is the common essence of reality?

I can understand that persons with a critical attitude would never take seriously a philosophy that is based on illusive suggestibility. I could have misled myself if suggestibility is only a side effect of something else, something which is consistent and real, that has universal meaning, but is not necessarily physical. Obviously, physics has inserted blind mechanical energy as content, but without conscious being the world becomes meaningless. Something could exist that I did not know, that could explain what is missing. I decided to wait as I might see it in a different light in a few years.

This insight was a sobering setback. A year passed, and I could not completely forget the missing factor and became convinced that somebody else with better information might eventually find it. The grandiose idea that mankind is close to the understanding of a single cause for the universe I still believed, and I was not the only one. Occasionally I asked myself, what is the key to understanding the hidden principle in nature to which suggestibility is only a part?

At least now I knew that one important factor was missing that regulates the content and meaning of suggestibility—the important second half that could solve the puzzle. It must exist somehow and I was disappointed and relieved at the same time, ready to forget the entire subject forever, knowing I had developed another theory that collapsed when confronted with reality like all the other philosophies that had become history.

About six months later, half in irony, I asked myself again what is a common thing other philosophers had already written about as the "order in nature"? According to major thinkers with a complete system, like Descartes or Spinoza, from Leibniz to Kant and Hegel, and to skeptical materialists like Hume, Spencer and James, it was either pure reason or practical reason that guided the world. To others like Schopenhauer, Nietzsche and Sartre, nature was irrational, compulsive and active, or driven by will and existential needs to dominate, which were innate to nature. To most other philosophers without a distinct system of their own, the order in nature was based on freedom, love, greed or wealth, things that do not apply to everything. There was always a cause, a deliberate intention, be it good or evil as a motive pointing to goals, something different in each case.

None of these ideas in particular seemed to fit one single idea in a word to go with suggestibility. But if all these ideas were put together, if that could be expressed in one single word, what might it be? I still did not comprehend what this one idea could mean and gave up.

In philosophy, one can find many reasonable answers to any question and their direct opposites as well, except certainty or proof that satisfies everyone. I didn't bother to ask theologians, because I knew their answer: It is the mystery of God's unquestionable wisdom. The problem even with answer is that many think the world is irrational. For simple folks there is only love, but this belief is unreal. And for the strong-willed, who believe in the will to assert with power—this is unsustainable. For instance, love excludes all the bad things it dislikes, which is typical of all opposites as good needs evil to give it meaning. The religions need God to be good and to blame the poor Devil for all the evil in the world. There has to be something that is both good and bad to blame for everything that exists, such as the god Shiva in Hinduism.

Let me now ask what "average" humans with common sense, who usually believe speculation is a waste of time, think? The majority of people in the world are preoccupied with themselves, and are self-driven to strive for material security, many knowing no limits in providing material security for a future that humans do not have. In their obsession for security, which we call greed, they gather wealth, but greed is not a true motive. People do not actually strive for material wealth, but for what it suggests—influence and importance. At second glance, to be important could fit with suggestion. This means to suggest that importance hides behind wealth, even behind knowledge, and certainly behind power. Since the earliest times people have always known that people are conceited and praised the powerful to flatter their egos, which is cunning and dishonest, but where does this leave religion? In fact, respect and recognition is what the powerful want and also what they reward. Singing praises is as old as time. Maidens blush when men praise their qualities even if they know it is flattery. Although we see through praise as a method to gain favors, indirectly we all need some confirmation of worth. Of course, everyone is different and special, and one has to know what has meaning for a person.

Yet one can also suggest negative qualities to degrade the worth of a person. Suggestion works as insult and can emotionally affect people. From psychology we already know about the unconscious need to feel important. All beings need to feel that they have worth for themselves—to be something. Yet that leaves out all material objects that many humans also love, like money or objects of art or anything else that exists. It suddenly occurred to me that the common denominator would be value. There is value to human emotions and also to material objects. The suggestion of value is the universal motive for anything and everything, receiving it and giving it, positive or negative made sense. This includes everything that exists and can exist because it has value for itself or it would not exist. Besides positive and negative applies to electrical energy, which we know is suggestible.

But I am getting ahead in the years of my personal development, which happened slowly. One summer, about thirty years ago, we rented a cottage on Lake Bernard in northern Ontario for our vacation. One morning I awoke early. Since it was not bright enough to go out, my mind just started to think about the missing part. The idea of value crossed my mind. Values are like everything else, dependant on their opposites, but it must be a single cause that inspires the universe. Suddenly I understood. It is not the separate values, but innate suggestibility to value which is a single emotion that ought to be the cause for the organization of worlds and the universe of worlds, because value evolves from suggested needs.

I got up with excitement and walked on the beach of Lake Bernard as the sun rose to a new day. That early morning, during a walk, the general idea for the Universal Principle came together. Why mention such personal details? Because personal emotions are important in an emotional-based reality. I was elated and knew that a great idea had come from my unconscious mind when I was relaxed and carefree. Just after awakening from sleep is usually when ideas come into my consciousness. I knew that value was the missing half, the one word that can mean all things, mental and physical, good and bad, positive and negative, including profits and losses. If nature is understood as a process in which values evolve and create worlds of meaning by suggestion, it can be understood as a single phrase: "Suggestibility to value"—a single emotion that can inspire small and large events. It means that evolution is a value-creating process inspired by innate suggestibility, and all aspects of reality are the result of it.

We are all suggestible to praise and admiration, because it confirms our value. Unconsciously, which means indirectly, it affects the organism and raises our spirits to generate positive energy. In the human experience, to be valued is the same as being loved. But value also has the inevitable opposite, worthlessness. If the suggestion turns negative, the energy diminishes and this can lead to disorganization. This change of mood from positive to negative means the essence of the universe is emotional, and this fact applies to electrical energy which is also compatible with quantum theory in physics. It must be universal and true for all organized things and beings. If the universe is organized by positive suggestions, this would also mean that negative suggestion disorganizes the universe, and affects things that comprise it. This is true, whether for atoms, stars or humans. There are no exceptions.

This is more complex still in the details, because we are all different—from the cells to the organization of the brain, which is also affected emotionally by the conjunction of the planets at the time of birth, and foremost by our personal experiences in life. Many suggestions can affect us unconsciously. Because everything is conscious, the fact that selectivity is relevant to identity points to a need for intelligence. In a physical sense, reality is a wave phenomenon, values before they manifest are ideas and ideas can change. But once physically manifested, ideas are difficult to change, except in a new organization. That is how change by evolution happens—by frequent reorganization and the occasional radical transformation of levels. This requires the death of the inflexible materializations.

From the general pattern of nature one can speculate as to why the universe exists. It is inspired from the general desire of countless masses of suggestible entities which existed from the beginning—prior to the creation of worlds in the absolute void. These suggestible entities, which are emotionally excitable, first became aroused to a frenzy by the suggestion to experience interesting events and to be something of meaning, and then congregated into massive, fiery galaxies that illuminated the darkness of the void with light. The galaxies separated into star systems with planets of different emotions. This separation continues in the further organization on the diverse planets into different experiences of reality, which can exist on frequencies which are invisible in our mode of perception. After cooling, physical entities organize further to become manifestations to experience—exciting events. That is what the excitement is about—to become something. Anything is more than nothing; a wave caused by emotion in the void is something.

On a planet with favorable conditions with a biosphere, like Earth, through gradual evolution into manifold experiences, self-conscious human beings seem to have reached the understanding why "they" exist, and why the universe exists, namely to experience interesting lives by participating in the world process one way or another. Compared to nothingness, any organized manifestation is already more than nothing.

I fully understand that persons who have their minds made up—believing that objective methods are required in order to find the truth—would be unimpressed by an emotionally based concept. Still we need not all think alike. However, mathematics may have a similarity to suggestibility in the external structure of levels, it fails to adjust when wrong factors are introduced. Whereas innate suggestibility rejects unwanted ideas after it has awakened from a temporary trance, proving it is conscious and selective in relevance to meaningful identity which a mathematical system is unable to do.

Suggestibility and value belong together as one combined experience, even if it takes two words to describe it. Knowing that suggestibility to value is the motive for anything and everything indirectly is the indirect principle of reality. The universe is therefore a colossal meaning creating process with one intent, which is to experience meaning where otherwise nothing would have been. It is a universe of special meaning for everyone and everything in their own individual ways.

Individual life begins with strong suggestion of value and levels off during life when interesting events lose value and the excitement gradually diminishes and the suggestion that holds the organization together loses strength to maintain order in the cell state. Physical structure becomes dysfunctional, collapses and disorganizes. Whether the conscious identity also disorganizes is another subject. From many factual experiences we must conclude that individual identity continues to exist as an idea, like an immortal soul, as an energy field with a value identity retained for itself. For how long a spirit exists is still speculation. With this general understanding of reality as a value-creating process, the meaning and purpose of the universe was answered for me. It was answered because my mind was saturated with the wish to understand. In fact, it came to my mind first as an idea, far from clear, then as a general understanding of a system that made sense. Many things that I had mentioned in the first chapter were then still unknown to me. The important thing was the concept, which I now had, to which more details still had to be added for most of my life. In fact, one is never finished.

Let me again return to that morning when I got up at dawn, to walk by myself along the beach on Lake Bernard. At that moment, I was emotionally elated and excited because this insight completed the Universal Principle as a functional system. It felt as if an illumination had occurred, as if a child had been born and entrusted to me—a child who was still helpless and in need of nursing, and I had been assigned to see it grow up until it could stand on its own. It felt like a commitment to see it through, no matter how long it takes.

That such insights and feelings happen to many other humans shows that unresolved issues which lie dormant for some time can organize knowledge and become conscious as a new idea. It implies that the subconscious mind processes information while we are doing other things in life. This unconscious computing of meaningful ideas has been described by others before, notably by the physicist Hermann von Helmholtz, who said such ideas occur to him occasionally and have three stages. The first is the saturation of the mind with data, the second, incubation for an unknown time, and the third is, eventually illumination when the idea has matured.

It is a myth that scientists are detached from emotions and use only experiments and mathematics to discover the laws of nature. Helmholtz was a scientist and Nobel Laureate who considered the unconscious as an important source of the creative process. Like all of us, scientists are suggestible humans first and usually start with an idea that suggests a possibility to be pursued. Then one begins to experiment and calculate in order to prove or test the idea. This apparently was also true in my case. Inspired by the expectation that a single principle should be the sole cause of the countless variations of things and beings in the universe, I presented a question to the unconscious and received an answer. Because I had collected and stored information over many years, it was therefore possible that I should also have an illumination. I call it a vision of reality.

It is also interesting that this answer came when I had almost given up and followed other interests, when the unconscious began to help me along. This happened when my mind was relaxed, while I was on a vacation without pressures and concerns on my mind.

Such ideas also take a direction of their own from what is expected. This happened when it became necessary that all things should have value. Positive and negative, and good or evil must all have a value. This is an issue where philosophers normally stopped. In the natural process of evolution by division and separation of values, estrangement resulted, creating a condition in which congenial values harmonize and incompatible values oppose each other. This is the reality of the world. Some values are emotionally set against each other, even violently opposed, yet most of the time coexist peacefully while still keeping apart from each other. The more intelligent cooperate to help each other just as humans do. The question is still open about which values are constructive or destructive, good or bad, and to whom. Nobody can be certain because ideas about values can change, and friends can become enemies and enemies become friends. This means change is frequently occurring. At least values always mean something—that is more than nothing.

One must admit, just to want to live at all in this demanding and stressful world, it is necessary one should be born with a strong conviction of self-worth and retain it throughout life in order to survive and to suggest enough value and meaning to continue the daily struggle. It is not just a struggle for physical survival. Life must have meaning and value beyond eating. It must contain interesting moments and experiences of joy. Existence is oriented to the future. Only when the future becomes hopeless will the inspiration and the suggestion of value break down. We call this a mental breakdown on the human level.

Anything that exists, even our meanest enemies, the most evil bacteria and viruses from which we die miserably, also have value for themselves, and need to regard themselves as important in order to keep on fighting against extermination. However bad their situation is, they must adjust unconsciously, to invent new improved strains to assert and defend their values, which may just be their purpose and joy to bring large beings down, to maintain a natural balance. There is unconscious intelligence involved.

This proves that in the final understanding of the value-creating process, good and evil are only perspectives of a given position in the scheme of things, in the role of nature. On the organic level there is the food chain. We all live off each other indirectly. In nature the beings, depending on each other to stay alive, are all interdependent. Nature is committing the world to experience meaning where otherwise there was nothingness. Just to eat means to terminate the lives of other beings. But eating is only part of life. To experience value and meaning in a human body is the purpose of the human experience.

I knew that I would have to agree with some of the ideas held by warmongers and hero-lovers like Heraclitus of Ephesus, Hobbs, Darwin and Nietzsche. It may be true that violence and suffering are part of the value-creating process and necessary in order to remind us to appreciate peace, health and the joy of life, and to hold onto it. It seems logical, that in order to value health, one has to experience how it feels to be sick. The constant possibility of death hanging over life is necessary to give value to life—to appreciate being alive. What these philosophers saw was true, but it is the dark side of life that must be there to remind us of the bright side, to remind us that with intelligent living, suffering can be reduced. With foresight and reason we can avoid war, avoid illness and live reasonable lives.

Therefore, social values as a particular development of human reason evolved from social cooperation to live in peace and to experience interesting lives. Morality is absent on less socially evolved levels. In general, humans are on a higher plain of social development, but in detail, there are still many relapses of desperate self-assertion, which often seem insane. Yes, that is how the world appears to function, unpredictable in detail, but understandable in principle. In general, we can predict normal human life spans, but in detail we can not. Probably this will have to remain so. If the universe is a *continuum infinitum*, it has to remain imperfect and incomplete to continue.

In nature the suggestion of pain, which is designed to warn us, is necessary to avoid injury. If we ignore warnings we suffer the consequences. The suggestion of painful consequences is usually enough to deter reasonable people from pursuing risky ventures. This implies that reason is dominant over blind action, at least in sane, reasonable persons. But unreasonable beings and angry individuals who are not deterred by reason, with their mind closed by fixed ideas, must expect to suffer pain, even death.

But one also cannot live without courage. One must face dangers and life without fear. Something is there in this fearlessness that transcends the reality of dangers and seems to push them back. Yet without some suffering life would be unreal. What seems a contradiction in our striving for perfection as an exercise in futility is actually necessary in a continuum based on imperfection and incompleteness.

That is also the reason why all theories of a final goal or clear purpose of evolution are unrealistic when compared with facts. Facts show changes in detail, but the principle itself is the same general process of repetition of reorganization whereby improvements are made which have their dark side as an opposite. With more industrial production we have increased pollution of the air and drinking water and so on. If, however, it should happen that all problems are solved, that particular world would have to terminate itself and start from the beginning. Perhaps that is what we see out in the universe, but it is not the whole universe, only worlds that end. Galaxies are reorganized eventually in the timeless infinity of eternal consciousness that never is not. The universe remains, only galaxies reorganize. The meaning of cosmology is to experience interesting events. That is basically what existence is all about for stars and man alike: A separate experience of meaning.

The expansion of consciousness and the development of reason are the directions taken by what seems a deliberately slow evolution, which obviously is favoring open-minded and flexible individuals. Intelligence is rewarded by nature, and by mankind through recognition and rewards, but money is not the main issue as it frequently lures people from a meaningful and interesting life. When Gottfried von Leibniz said, "This is the best of all possible worlds," it was obviously meant as a self-perpetuating system. Effort and commitment are necessary to appreciate life. We are not actually free, but must live in the world as it is, wisely designed to continue forever, a forever that is only conceivable as a pluralistic consciousness that existed before worlds ever had organized and will exist when they disorganize.

My understanding of nature is based on the real existence of innate suggestibility, a spiritual essence of which we all are replicas. We are sparks of the emotional fire that burns in stars and galaxies, part of the suggestible psyche that created itself worlds in order to experience meaning and value. One can call the universal spirit God, but then we are like God. We are all living with an innate suggestibility to value, having evolved to understand the meaning of being. That the many religions were steps on the way to self-understanding is obvious. They are all the cultural heritage of mankind, and of undeniable importance in any understanding of human development.

Assuming that good and bad, like positive and negative, are created by innate suggestibility, and that nothing is absolutely certain or indispensable, then the only permanence is suggestible consciousness. All else is infrastructure to help in the experience meaning. If the purpose of the entire process of the world is to experience interesting lives, then suffering is not a goal, merely a threat to keep us on line to value being alive.

The expansion of consciousness to experience interesting lives is already the direction in which more advanced individuals within mankind are progressing. The information age has already begun with wider literacy among mankind. We are widening our horizons by experiencing interesting events, and entertainment keeps us emotionally inspired. Just for the record, 2,500 years ago Heraclitus said, "In dreams we are participating in the affairs of the world." Now we participate in an awake state in the affairs of the universe through the electronic information media—not only while sleeping do we consciously hunger for entertainment and knowledge. We consciously pursue what we already did unconsciously in sleep, but with one reservation—dreams are transcendental, time and space are unimportant, and only the experience counts.

The model of reality I herewith describe is only a metaphor, based on innate suggestibility to value, which in thought transcends the separations of value and can connect to other dimensions that already exist in order to understand the meaning of the whole. Human evolution has progressed from violent intolerance to coexistence in a precarious peace forced on us by nuclear technology. The philosophy here explained suggests a new model of reality that has a larger vision than the previous models. It certainly extends beyond the narrow model that physics has proposed of a strictly mechanical universe of matter.

In ancient Greece, during that period of human history when western philosophy began, when "man was the measure for all things," the temple in Delphi carried the inscription, "Know thyself and you will know the universe and understand the gods." This understanding comes from studying the inner self and studying the psyche, and would make psychology the ultimate science.

A similar common sense assumption had occurred in ancient India where seers compiled the *Rig Veda*, which says, "Nobody knows who made this world, not even the gods, they came later."

And the oneness of all beings was expressed in the statement, "You are that." This, however, is not to be taken directly. Since there are countless and deep separations between things and beings—separations that are real for us—it means that we evolved from one common source, but like to be different. There are the physical barriers that only consciousness can transcend. Ancient Hindu and Buddhist philosophy is based on compassion with human suffering and the wish to break out of the illusion created by the "mask of matter," and to escape from this world. But to go where? Back to utter nothingness, to stillness?

Such negative perceptions are implying that evolution caused more suffering than interesting lives. It actually calls the effort of the value creating process meaningless and in vain. This is a negative position that obviously is not shared by the psyche of the universe. From a positive view of the value-creating process and of evolution, it follows that with reason there is meaning in life. With an intelligent understanding of the intent of nature we can enjoy living and participating in interesting experiences. As an old Chinese proverb correctly confirms: "May you live in interesting times."

The world has meaning, but that it takes continuing effort to live implies that life is meant as a challenge. We are meant to act, to work and to create values, but in balance with nature. We must not ignore the signals from nature that industrialization is making organic processes difficult to regenerate. This in part is also caused by overcrowding by human beings of this limited biosphere, thereby endangering the balance of nature. We now became responsible, at least for the infrastructure on our planet in which we need to maintain a natural balance to continue to exist as physical beings.

Occasionally, I have had moments of doubt about the value of this philosophy. Can it take the mystery out of existence? Yet living in a technological society, the mystery has already been purged by materialism that leads to far worse, to disillusioning nihilism. To think of the purpose of the world as a meaning-creating process is a far better scenario than the meaningless universe of scientific materialism, which takes the physical infrastructure as being more important than the conscious experience of being alive.

3

Identity and Fixation

AFTER THE EXCITEMENT FROM THE DISCOVERY of a single principle which repeats itself everywhere had gradually subsided and the family vacation was over, my mind returned to practical reality, namely how does this principle fit into current theory and compare to centuries of philosophy? It was obvious that I must see it from other perspectives that reflect the presence of innate suggestibility as real. The few intellectuals I knew were academically trained and followed the mainstream of scientific materialism. They showed little interest in hypnotism or transcendental idealism, even less in extrasensory perception or psychokinesis. The only connection I had to information was through books that were strictly materialistic or that concluded that reality is a contradiction of facts. Added to the confusion was more data about mysterious phenomena presumed to be psychic experiences, which imply that the universe is born from primal chaos by accidents and chance—the New Age philosophy.

My intention was to prove the opposite—that a clear principle of differentiation and organization exists based on innate suggestibility to value from the beginning, which required unconscious orientation to ideas to explain the meaningful reality. I asked myself the basic question, "Who are we?" This question has been asked by thinking man since thinking began. The obvious answer was that we are emotional and suggestible beings. One advantage we have over other species in nature is greater consciousness; we can think and develop a string of thoughts. Innate suggestibility to value is universal and basic to all—to organize sense impression into ideas and ideas into concepts that are uniquely human qualities. A typical explanation from scientific materialists is that the human mind is no more than electrical activity in the brain and we delude ourselves that it has meaning. It is correct that thought is accompanied by electrical activity when we think. And of this activity nothing can be known. What the thoughts mean is clear, although electrical brain waves cannot be translated into words. One still has to ask people what they are thinking. The orientation of life is consciousness and it organizes waves into meaning. Therefore thoughts have meaning and make sense of the world. The world is organized in a meaningful way—not accidental or meaningless.

Unless one is programmed to think in mechanical concepts, to exclude the existence of consciousness in reality, then one is living in an unreal material delusion. I just wonder how such persons think, if they think at all.

The history of philosophy is deeply involved with questions of identity, in particular whether we are automatons or free to choose, and in fact, whether we actually exist and the world is not only an illusion. Modern philosophy made a fresh start with René Descartes when he concluded, "I know that I exist because I think." But this answer leaves many details out. Most beings in nature obviously do not think, or at least very little, but still exist and function. Even humans in the first three years of life act without thought. And what may be called rational thinking starts at about twelve years of age. Even animals and plants are not automatons in a mechanical sense, because they can adjust to changing conditions since they are conscious and suggestible to value. They instinctively know what has value for their particular organism and what is dangerous for them. This is a limited range of consciousness, enough for a particular design in the menagerie of lower consciousness that is mainly concerned with eating and living.

If plants and animals, even human children do not think, what makes them behave in meaningful ways? The answer is their unconscious, which is also suggestible to value but in a slower mode. It is the unconscious that apparently has organized the world into separate entities, and uses the conscious level as outer orientation in the world, which it does through different senses of perception, only provided as far as is necessary. The differences in the senses determines what they perceive as world and sets the limits of the role that a species can play in the system of order. Nature appears to be organized on countless levels depending on each other. The details of how conscious orientation works remain unconscious in order not to be confused by the external world. In fact, the internal organization is better left alone or we have hypochondria. In spite of considerable knowledge, our understanding of our internal functions as explained by science is only established on a chemical level and explained as mechanical functions, without understanding any of the underlying ideas. Yet the understanding of ideas, of their suggested meaning, is of importance. It is a fact that the idea of individuals of their self-suggested value is not unlimited; this ideal can be disturbed. Criticism and negative suggestion can effect the ideal of organization, and can also effect physical growth and dissolve individual identity by confusing the mind.

Among reasoning beings such as man, many can think beyond their own narrow needs. To them, the larger meaning of the world is slowly dawning as they become conscious that existence is not a mechanical process, nor are the beings in the universe totally fixed for long. Meaningful changes are frequently occurring as feedback, in response to unconscious suggestions.

Probably such a feedback occurred to me that morning on a vacation thirty years ago as I contemplated the first cause of the universe. I understood that suggestibility to value is the first cause and that ideas evolve gradually. But all this activity involves effort, and it requires inspiration and must create the energy to transform from nothingness a phenomena that suggests tangible reality. Suggestion of value, as inspiration, must itself therefore provide the energy necessary to create worlds. Energy must be created to meet the demand needed by innate suggestibility—namely the emotional excitement which I then felt.

Let us assume basic pluralistic consciousness exists in a state of limbo. From calculations made by physicists, we could think of quanta existing in fields of infinite numbers of invisible dimensionless entities with a suggestible psyche that can be inspired to congregate into massive spinning clusters we call galaxies, excited by the idea to participate in the creation of interesting worlds for the purpose of experiencing meaning. The emotions of the masses of these suggestible entities mutually excite each other in a frenzy of emotions that generate the heat and ignite the fire that illuminate the darkness of the void as the light which is conscious in essence. The result is vast, cosmic energy, inconceivably greater and the birth of the universe of galaxies, stars and planets—each is a part of the grand excitable psyche. When on the surface of planets like Earth, similar emotional beings evolve, among them self-conscious humans who begin to comprehend the purpose of the universe. We know that we are a spark of the grand cosmic conflagration that illuminates our world, inspired by the suggestion of value which creates the energy by ideas.

This scenario is exactly the opposite of the vision proposed by scientific materialism. The materialists conceive the universe to be created from the explosion of super concentrated matter forced together by gravity. It is a material universe where emotional beings, let alone conscious humans, cannot be integrated, but have been added on as a freak accident, a separate "anthropological principle" parallel to the physical principle of reality. This already creates two principle which is contradicting the idea of a common essence. Physicists are somehow transfixed by a constancy of mass that cannot grow or diminish. That keeps them from seeing that we are emotional, suggestible entities, part of a psychic phenomenon that is able to raise energy if inspired.

The "psychic model," or, as it also has been called spiritual, has a clear motive that repeats itself on all levels and beings—to create interesting worlds, compared to a lack of motive and a meaningless chaos that mathematics implies with fixed, immutable laws in a physical model of a material universe.

In the living reality organized by ideas, suggestion of value is the common energy that drives each individual entity, and is the only possible way to explain the organization of cells into highly organized beings, as temporal manifestation, who after a brief experience in the material world disorganize physically, but do not necessarily suffer a loss of consciousness and memory of their identity. In fact, there is a lot of proof that the spiritual essence of personal identity continues to other levels and modes of being. The suggestion of reality provides an explanation as to why souls, as spirits, continue to experience meaning in the value-creating process since the intent is to experience interesting lives, and changing to another manifest identity is consistent with that purpose. But death of consciousness would end the continuum—the continuum of which we are a part, following the same principle as the universe in general. This termination of consciousness is most unlikely. Further explanation as continuation is more likely.

In the past, the idea of self-creation of the world was never seriously considered because the beings in nature, including individual men, are far too ignorant to create complex beings like themselves, and are only following their instincts in providing the possibility for it to happen. As a human, one can opt out of the suggestion to procreate. This is a freedom of choice that we have, but to organize the cells of a living being is outside our ken. At least we understand we are only links in an evolving chain of human consciousness, which again is a part of universal consciousness which creates its own energy from emotional excitement.

It is conceivable that the memory of the sequence of organization is present in the unconscious, out of which all things have organized in a slow process of evolution, implying a continuation of the idea of the self through the memory of the organization that is unconscious, to keep separate of the current life. Can we assume that the unconscious must have the knowledge of how to organize the physical body, but we do not consciously remember? Such massive memory of the details of organization would possibly be a burden on our outside orientation and distract us from actually living. Besides, we are prone to make rushed decisions on short-range needs, that would affect the general security for future generations. Such mistakes may have happened to other species when they fixed themselves by developing armors, large teeth, hooves and horns, which became a burden, while man developed intelligence and cleverness instead. Once a radical design is physically manifested, a species is fixed. If we had hooves instead of hands, what would we do?

Apparently the human unconscious was careful in the organization of our body. By choosing expansion of consciousness towards intelligence, it avoided physical fixation in a dead-end line. With this open line we became rational, thinking beings who depend on our intelligence to survive. Consequently, our conscious mind is our greatest asset, and it must remain clear and open to changing situations. In fact, every situation demands individual consideration, and fixed, mechanical laws, even if sciences likes to believe it, are misleading and should be avoided. As it is, every new life begins fresh with new energy to grow—a wise idea that nature has provided by forgetting the past. To forget and thereby forgive what ever happened in the past is the deep wisdom in the design of nature from which we should have learned something.

Because the organization is unconscious, other individual characteristics of personality are also unconscious suggestions. To eat is a natural compulsion; so is the suggestion to procreate. These are strong suggestions that we share with all organic life, but we have evolved to a level of consciousness where we can control these impulses. We can delay these impulses until the opportunity is favorable and socially acceptable. For example, the control of the sex drive is enforced by law, and the use of unhealthy drugs is restricted, but the control for overeating and the indulgence of other physical pleasures is left as a personal choice.

Freedom of choice demands self-control, but many persons are out of control in certain compulsions and literally try to destroy their health. Yet we all seem to be dominated by unconscious suggestions one way or another, because we have to orient to current needs, which means to live by reason that implies a struggle with oneself.

We identify with many unconscious suggestions and adopt them as values as part of our individual identity. Most common are organic suggestions, next are personal experiences, and then acquired knowledge to assume an identity.

To identify persons as types has been tried before, but it is very complex and subject to uncertainty. In earlier ages, the stars were considered to have suggestive influences on people, and astrology was a science, using calculations. Then, race and ethnic factors were considered important, gender was a stereotype. Even the clothes one wears, the family one comes from, and education have become important. But all these factors are secondary to intelligence. With intelligence, one stands above such limiting suggestions, sees through them, and knows what is important. That one still cannot convince others who believe in classifying and in class distinction is true. Life is challenged by the obstacles which one encounters and overcomes.

The belief in fixed values which are stable and dependable to calculate with and formulate laws is the essence of scientific materialism. But this belief is based on observing the lower levels of nature—in fact only the materials level is stable, and that not over longer time ranges. Nothing is permanent, everything transforms at different speeds, and grows as well.

Consciousness is not restricted to organic life. The material level is also conscious, but introverted and nearly autistic and reacts to emotions. For example, complex machines are assembled from parts in a meaningful organization to function by mechanical laws, but the innate consciousness of the materials remains separate—almost dead.

I am not assuming that materials have a similar level of consciousness to ours, only that they are also emotional and suggestible, which indirectly means conscious. They have fixed values and only reluctantly allow themselves to be organized by man into larger designs. When atoms and molecules evolve within organic systems this seems to suggest value and inspire their cooperation, but only as long as the suggestion is convincing they are a willing part of a meaningful organization. This is rational. Human beings also would leave an organization that is failing. This is based on the premise that suggestion of values inspires and maintains everything that exists as long as it can suggest value and meaning to participate. Suggestion of value is creative and destructive, depending on whether it is positively or negatively inspired.

Our human body is no exception. Comprised of vast quantities of atoms and molecules in our cells, the design is apparently stored in the genes, the collective experience of countless living entities from whose knowledge our bodies are organized. What is the suggestion of value that keeps the organization together? The promise of interesting experiences in the organization, for which one has to give up some of the freedom as a separate entity. But it must be a meaningful experience. If a life experience is depressing, the interest in the organization is lost and the union will malfunction. This is also the practical truth of human existence. If you let your self-assurance and your positive outlook decline, so do your health and life decrease in value.

Therefore, the basic meaning of being is to maintain a feeling of worth, to be something of importance for oneself. Psychology and medicine confirm this fact. The world is the interaction of organized entities, but it is not a Darwinian struggle of survival, rather an interdependent process in which one tries to coexist. Man developed foresight and understanding by which he could avoid many dangerous situations before harm and suffering could take place. Instead of the strongest and most ruthless, the best organized species became dominant in the world. The history of nature proves it and the history of civilization confirms it. We were loyal to the tribe and united we became the dominant species.

In general we like to be fair and allow others to live, which is not a Darwinian deduction of reality and not naive either.

Let me return to the unconscious since many scientists officially are against making presumptions. I am trying to see it from their perspective. If one were to believe chemists that experiments proved that chemicals react mechanically because they can show it happen, this an external observation. They cannot know whether it was emotional, a positive or negative reaction. All they can do write it as a formula, such as $H + O_2$ combine as water, because water has more value therefore it is a positive synthesis and occurs by itself. That this transformation has meaning and purpose for other living organisms implies it is of value to our planet. Water provides the basis for most living beings to exist and evolve—which is essential to further evolution.

Another example is a chemical combination of drugs that tested useful to cure overactive imaginations and thereby lowered the severity of mental illness. This can confirm that the brain is also a chemical laboratory, at a level where psychology failed to achieve a breakthrough. The barrier that protects identity, in my understanding, is a level of consciousness. But chemists have nothing to do with such imagery. It is simply proof that the brain is a chemical laboratory and conscious ideas are unnecessary. Instead it is chemistry that effects everything, from the body to the mind. Therefore, if a person becomes mentally deranged, it is from a chemical imbalance in the brain—to correct the imbalance is the cure. Yet the reverse is more likely correct. The emotional mind produces the chemical imbalance and all chemicals in the body are also conscious, as are the cells. That is not even a question, because the entire organization of nature is meaningful—intentionally organized. But in this particular case it becomes clear that the drugs given to mentally unbalanced individuals only lower the emotional frustration and open a door to communicate. Now at least one can talk with these persons about their fixed ideas. If one can convince them that their ideas are wrong and they see it themselves, they are cured. If they remain without counselling, or cannot be convinced, they have to remain dependent on drugs until they pass away. This proves that the mind is creating everything, chemicals included. Mental illness is just manifestations of emotions frustrated by life.

We humans are on a higher stage of identification with self-suggested, fixed ideas that can only be changed by helping a person to understand why he or she is ill, and if that person understands this clearly as cause, means that indirect self-suggestion can break the fixation.

This is where the suggestion is effective, and indirectly it must restore the self-worth of the individual to trust himself. Ultimately, this happens through self-suggestion although innate suggestibility to value is the explanation of the creation of chemicals in the body, it can turn against itself. Chemicals that usually help can cause an allergic reaction to poison the organism, depending on the emotional intentions of the individual unconscious. Like discontentment with certain aspects of a life, when life is perceived as worthless, the self gives up.

The fact that we are all differently endowed genetically from birth implies that nature has deliberately suggested different values for each individual to identify with—to guarantee interesting diversity in the world by experience exciting individual lives.

The entire cosmology of scientific materialism could be challenged by this single principle: If the fire and light in stars and galaxies were emotional excitement in anticipation of the creation and experience of interesting worlds caused by the innate suggestibility, then excitement is a positive experience. Although on our level this early stage in evolution seems a violent process, on the cosmic level it could mean ecstatic pleasure that "generates" the wave energy necessary to organize worlds—countless worlds of various meanings—which constitute the universe.

This conclusion is a consequent deduction from many ideas and the assumption that unconscious reason is influencing the direction of organization in general, one assumption that science resents as spiritualism. But it is nothing like the Holy Spirit in religion since consciousness is innate in all beings as suggestibility of value, and we are all a part of it. Creative evolution is imperfect and searching for meaning, and so are all its individual expressions. There is no need to call the universe or any of its beings divine, because many mistakes confirm that nature is still learning as it expands in consciousness and evolves by its own energy.

From the idea of a searching evolution, it follows that all things originate from one source or essence, and to this I wish to add, move in the general direction to create interesting worlds to experience exciting events and changes. Nature is not just explained with random energy, but is rather a meaningful phenomenon of complex ideas that produce a multiplicity of things and beings that actually require consciousness to perceive that it exists.

A reality check

To check for reality again to satisfy my own peace of mind, I took that second critical look at how real people behave and what they identify with, thinking this should test the Universal Principle, and make it or break it. What I found out was that the intent in people to create values and change the world was weakly hidden from others. The majority of people were careful and skeptical persons, without a wish to create anything too different and outstanding. In fact, most persons were conservative, disliking change, and preferred security over taking risks. It seemed as if normal people just want to live without problems and obstacles, and be like other average persons—feeling secure in being part of the mainstream. The greater majority of average people want to be normal. Yet can one believe people when their reason tells them what to say and do, to be accepted as normal, or will they act according to what their unconscious wants them to do?

A possible self-deception concerns me in particular. Am I misleading myself in thinking that I have discovered the psychic principle of an emotional universe, when it is no more than a human self-projection only of one individual? Skepticism has its place, although I am very concerned that I remain real and refer to the actual world and real people whenever I can. I would be grateful to be shown where I have erred. Yet it cannot be an error in details—that does not count. I know that details are changing, and of many things I am ignorant. Effective criticism has to challenge the principle idea, concerning organization and the reality of innate suggestibility. I refuse to look at a mechanical explanation of the universe.

Uncertainty is elementary in understanding human existence and nature. It is the reason why most prefer security in groups. Humans seek the company of each other for support, and in groups may take risks when they find that many others will join them. This fact demanded a slight change of the interpretation of reality. This is not the straightforward value-creating process that I thought the world is, up to now, but it is held back and slowed by security considerations of the majority, and by loyalty to accepted, traditional values.

It is, therefore, delayed evolution in steps, also in steps backward if value is perceived. This also happens to quantum jumps, which is more like the reality out there in the world. Evolution in principle is evident but in detail is complex to define. Human evolution is also complex in detail, because the human body has a few changes in design that are more like planned changes—deliberate delays, even retardation.

Most humans are insecure and fear that change will demand adjustments and this is exactly why most people belong to the conservative majority—to be on the safe side. In regards to the value-creating process, it is still there. Even modest persons create value. The average person also takes chances, even if only in little things for practical living. I therefore concluded that if the real people that I personally knew were typical of the majority, then it is only a narrow minority that openly expresses the intent to change the world and challenge established values, while the majority follows established values and conventions.

One could take the average individual as prudent or overprotective. As long as one agrees that they are conservative, this means that established values are difficult to change. People fight for their identity, whatever it might be, often with heroic and fanatic inspiration, either for something or against it as long as there is a cause which suggests value and meaning. Whatever idea it might be that sets them on fire it has to suggest value at that moment to create the energy necessary to stand against something.

This belated insight actually made me change my understanding of the common principle in regard to the value-creating process. It is also indirect. I now see a second side, it is held up by the necessity of temporarily establishing an identity because this is what human life actually means—to become something that exists for a while in physical form, a manifested self, a fixed consciousness in the stream of time. It is like standing in a river while seeing the changes float by. In the conservative mainstream of mankind, stability and constancy are expressed. Mankind has a separate evolution within the larger evolution to establish and secure a temporal identity—long enough to become an individual being that is able to think for itself and to experience meaning—before life is over and one goes away, perhaps to another deployment where an identity is a requirement to be admitted.

This delayed pattern is also reflected in the intent of all living beings to prolong existence and to cling to life because it has value. One is born into this world, in our case as humans, and has manifested itself as a physically predetermined person. The content of the mind, however, is shaped by experiences and the influence of many suggestions of value—depending on what values are relevant to the individual. One acquires a set of individual values from the impressions one experiences in life. If that environment is narrow, one has narrow values. If one grows up in a wider horizon, one is more open to alternative life-styles. Here we have an answer why narrow-minded people are intolerant. It is because of lack of exposure to other values. In the process of establishing an identity as an independent individual, values become fixed. People are born with the will to accept themselves, and thereby seem intolerant, regardless of ignorance. I am what I am, loyal to myself. Differentiation is the normal intent of evolution in nature, but for civilized persons cooperation with others is essential. This means that individuals must evolve to be tolerant and independent beings to fit into human society.

On the material level, protection of identity has resulted in withdrawal to introversion, to a state of autism. Atoms also have closed themselves off into a narrow range of values as identity. The nucleus is the stable identity that is protected from influence, only the electrons that spin around the nucleus are willing to take chances. In their adventures, electrons organize into larger designs. Normally things and beings fix themselves before they venture out to take chances, to make sure they keep what they have. This securing and fixing is everywhere, even in electronic computers. Fixing is protection—the reasonable explanation of security, which nature has followed. Flexible, mobile beings have organized into many species—trying to avoid useless conflict if they can, by evolving in alternative ways to survive. Yet, in many circumstances, such as overcrowding, where food and space are limited, conflicts are bound to happen. The problems caused by overpopulation are counterproductive to the growing trend of tolerance among civilized people, and the general expectation that the evolution of reason will make life less violent in the future is thereby in question. The way to avoid conflict is by self-control. With increased understanding, violence and suffering can be avoided. We shall see what the future brings.

Is there another explanation for the conflict of values? The growth of reason in the world was observed by idealists such as Georg Hegel, who thought it would lead to further civilization of mankind. But this belief was rejected by the emergence of existentialism, destroying all notions for reason in nature other than blind will. Schopenhauer and Sartre called the world a savage and irrational experience—not worth the effort. Kierkegaard rejected reason from the point that it destroys the religious mystery, and Nietzsche proposed to say "yes" to the turmoil and conflict, which forces man higher in creative evolution by participation in the process. What can one add to such divergent conclusions of Europe's philosophers? If nature seems like that, then we have to live with all possibilities. But can we use reason to foresee and avoid what might happen—staying in the middle between extremes, and when change occurs, it will be peaceful.

Of course, there were also philosophers with a pragmatic view of life, saying, take what you can for yourself and forget what happens later. But pragmatism was only practical when unlimited space was available—when America was still relatively unpopulated. All of the above are partly correct, but Hegel was nearest to the truth. He said that the world is rational. He did not know that even the unconscious follows reasons and orients itself by value. He did not know about suggestibility of value as reason. Yet he also tried to develop a logic and was wrong in assuming the existence of absolute reason. We are all suggestible to changing values and are emotional by nature. In emotional situations, humans can become unpredictable and unreasonable, particular when the unconscious takes command and acts like the rest of nature from which we descended—unpredictable animals.

There is no issue, however. These are two stages of development. The organic unconscious is impulsive and the consciousness of civilized humans has become more rational.

This is a case of a new level of consciousness evolving on top of an older organic consciousness. This older consciousness becomes unconscious in order not to confuse the clearer level of a more rational consciousness. Because confusion does happen occasionally as break-in of unconscious content, we refer to it as mental breakdown or the confusion of peoples' minds and identities. At this point the skeptics would object, saying that there is no evidence of other levels of consciousness. When one dies, that ends all awareness. This is another subject that will be dealt with later.

From a point of organization and concerning the human species as separate procreation of new individuals, this repeats the evolution of organic life as reorganization from sperm and egg cells by cell division. We also know how consciousness evolves because it happened to us. Self-consciousness begins to emerge at about age two when language develops. We know the mind level evolves on top of organic processes, because at that stage, we can remember incidents. We also know that the physical body can live without conscious orientation in coma. The question is: Can the mind also live without a physical body? One other factor usually left out or given low priority is extrasensory experiences and the influence of the unconscious. This provides additional information by telling us consciousness is universal, but physical structures are temporal.

In ancient history, "providence" was given credit for the outcome of many events. Herodotus mentioned it, Hegel agreed with it and so did the psychologist Carl Jung, who called providence the mysterious synchronism of events. These thinkers studied history long enough to see patterns emerging. Our normal reasoning can also understand it. Without some guidance from a distance, the complex organization of our world and of the conscious mind is inconceivable. In a way, we humans lived with uncertainty for ages and one more mystery cannot be news. Uncertainty is the spice of life. If everything were predictable, life would be a forgone conclusion and uninteresting at that.

From my general impression of people, it appears that the majority are convinced their values are worth fighting for and are willing to defend their identity and to resist being changed by others.

Delaying Change

The conservative majority should have a stabilizing influence by slowing progress and change. Without the establishment of a fixed identity, nothing could ever remain stable and acquire any status of excellence, including culture, literature or knowledge. Stability is as important for values as change is for progress.

A strictly forward-directed development could speed through life in a few years, but we would not be able to reach maturity and reason, therefore the need for delays. This allows me to speculate: Could the purpose of life serve to acquire a conscious identity through the physical body and experience—a fixed point of reference which on the next dimension gives one an identity?

In nature, progress advances in halting steps. After something of value is created, it is secured by fixing it. This means evolution advances in steps from creation to fixing, and what is has created is maintained for a while to experience it before a new idea is pursued. Thereby, the separation and reorganization continues, resulting in temporal manifestations of value.

As delayed advance, this is how organization and evolution make sense. Everything that appears on the scene must establish a separate identity and protect its value or it would never become real and that not for long. Once established, then it can venture to expand further. In this normal sequence of becoming, the important emphasis is the protection of value first and the creation of new values next in a chain of progression.

The general impression of the value-creating process that I held for a few years as straight progress—where fixed ideas are only a hindrance—was wrong and unreal. Therefore, my philosophy demanded a major adjustment, because fixation is not a hindrance, but a necessary factor in the value-creating process. The material level is evidence of it and so is the majority of conservative people. From this impression alone, nature could be compared to a value-conserving process that strives for stability, counteracting an otherwise speeding-away evolution. I knew that I had become older and wiser, seeing two sides to everything.

This reminds me of the relative stability of the material level, which once manifested, wants most to remain what it is. On the material level where protection of identity has resulted in fixed ideas of value and elimination of all communication, the entities remain disinterested in others, and exist in their own suggested glory. One has to recall one important point learned from hypnosis—namely the protection of identity against unwanted suggestions. Protection is the second and most obvious factor in the entire phenomenon of reality, defining evolution. The aspect of securing values is most important in a system created by suggestion, to protect itself against confusion from unwanted suggestion. Contradictory suggestion could lead to the collapse of the entire system in total confusion if the protective separation were to fail, particularly in so fragile an organization as an individual human.

Indeed, mental illness is a weakness in separation, resulting in confusion of identity. The system of reality, however, itself does not collapse. It is confined to the separate psyche of individuals, which means this is a pluralistic universe of uncountable souls, where the reorganization of single individuals is the normal way to renewal. This would imply that the universe is constantly changing and replacing itself—that the current theory of physics of a singular big bang is incompatible with observed reality.

The value-creating process moves in steps, alternating between creative organization and fixing what was created. Nature, then, is unlike the even flow which the word "evolution" implies, but rather has the character of change occurring in meaningful jumps in values, complete and viable, expressing a qualitative change. A change that occurs in distinct units or bundles and jumps suddenly higher or lower. Even flow is not real. It changes in value leaps to other levels and modes, like quantum theory implies but as distinct quality.

As humans we are part of this process, and we can take the fact that our own body has remained unchanged as confirmation of complete jumps instead of gradual change. Changes do not occur unless necessary. How else can one know how the internal process works but from one's own experience? We can influence ourselves by self-suggestion, but it works indirectly, as with anorexia. If one is consumed by a single idea. Accepted by the unconscious, once begun it is difficult to reverse.

Creative thoughts happen as suggestion of value in the subconscious mind. In a relaxed state we are able to transcend our individual borders, at least in thought, and seem to have access to information-sharing that connects us to a pool of ideas, which Carl Jung calls the "collective unconscious" and Socrates refers to as "eternally existing" ideas. But this source is careful and reluctant to hand out information and does it only after one shows serious interest. Because many were seriously interested in technical ideas, the result was the spread of inventive ideas that brought technical innovations. If the world is a place to experience meaning, to mechanize everything to make life easy, many are deprived now of a purpose and it will be worse for the next generation. Most people will be deprived of serious commitment and will feel even more useless than many people do now. Indeed, to make life easy does little to make one feel meaningful. Progress is not as useful as many think.

We can also think in terms of physical values. In the electromagnetic universe only positive and negative forces exist, which seem like pro and con, yes and no, right or wrong. The same basic emotions are also displayed in the human psyche as a positive or negative attitude towards life, which on occasion can reveal whether we wish to live or want to die. It is from the simple that the complex has diversified. I recognize the advantage of abstract mathematical thinking, but must point out that even computers are based on the yes or no, positive or negative system, which is actually emotional and suggestible if one includes all forms of conscious experience. The restriction of speculation to physical energy implies a limited concept on which scientific materialism is based, a one-sided abstraction of reality that has lost sight of the meaning of being and eventually will have to adjust to a larger principle of reality. That reality provides an increase of energy as the opposite of entrophy, refuting thermodynamics.

With the similarity of positive and negative emotions in energy and in human emotions, the possibility of unification is open, leading to a common principle in which quantum wave theory is made easier to comprehend. This does not mean the model change is soon going to happen. Alfred North-Whitehead, a defender of scientific materialism, referred to Plato's idealism and the presence of abstract ideas in the world as real. But he insisted that the universe is still in essence, material, and basic matter is without consciousness—that only the upper level has consciousness. This only makes a division where none exists, to salvage an outdated model. Consciousness as innate suggestibility runs through all levels without exceptions and confirms that one single motive explains everything, from emotional wave to emotional man who can generate emotional energy.

With the event of quantum mechanics, consciousness slipped quietly into matter—at least some physicists now want to think that the universe arose out of chaos to reach consciousness in man. But this is not enough to explain meaningful organization from the wave up, by innate consciousness which is suggestible to value. Actually, consciousness is a very old word for innate suggestibility to value that can be understood as highly emotional and excitable. A new model of the universe is still required, and materialism is not it. The human psyche is an example of the whole, a repetition of the principle in a nut shell. It is, therefore, on our level that suggestibility gives the best example of its existence. The physical and chemical levels are also suggestible, but compared to our range of awareness, must seem inconclusive.

Philosophy, long before science, perceived reality as a phenomenon of ideas, and a new school of scientifically oriented philosophy, calling itself phenomenology, emerged to integrate all knowledge. This happened with the support of many physicists who investigate reality "objectively" by reduction from the largest to the smallest, and from the top down. This scientific phenomenology analyzed the process of creation to the last detail. But its leader, Edmund Husserl, after several decades of investigation, found that the pattern of evidence turned in support of idealism. He admitted the need for unconscious reason to make the world functional. There it ended, but it did not convince ingrained materialists who want nothing to do with any trace of "intentionality" in the universe.

Husserl had never found a mechanical alternative for consciousness. What philosophers from ancient India to the industrial West have finally agreed on is the need of the prior existence of intentionality in the void. In fact, the unconscious is more involved than previously expected, particularly if the range of the unconscious mind is probed.

With this fact the important subjects of psychology and psychiatry make the subject very interesting. With it, the connection between dreaming and neurosis comes into focus, revealing the influence on organic and mental processes by suggestibility to restore health indirectly.

Sigmund Freud was the main originator of the psychology of the unconscious, yet he seemed to have become more deeply involved in the complexity of the psyche than necessary (which scared him) by the introduction of hypnosis, which he gave up for psychoanalysis. A classic example of trying to avoid disturbing the identity of the individual by the use of hypnotherapy is found in a case reported in Germany. A female patient's problems first started when she could not walk because of paralysis in her legs. This paralysis gradually spread to the arms and upper body, at which point chemically based medicine gave up, and she was sent to a psychologist for treatment, since nothing else helped. Because the patient was sane and could still reason, this was a case of neurosis, not mental illness. Neurosis only affects the body, not the mind.

This is when hypnotherapy was started with this patient. When questioned in trance, the lady revealed she had a deep-seated wish to cause her own death, to prove to her father that he was responsible for her suffering and that he should feel guilty for her death because of an "injustice" done to her by him. While she was being questioned, the impression was given that the subconscious answered as if it were a third person talking. The doctor, obviously careful to avoid further opening of wounds in the identity of the patient, did not ask about the details of the injustice but tried to reason with the unconscious. He tried to convince the third person's alter ego that her father was only human and should be forgiven. And he also made a deal with that other person in her that he would stop probing further with hypnosis, if in turn her health were restored. This then actually happened: the paralysis reversed itself. It had already become life-threatening, but within weeks complete mobility was restored and she could return to work.

My understanding is that her anger had caused the brain to produce a chemical to poison the body and, after being convinced to forgive, produced the reverse chemical to heal itself. This was clearly a self-suggested paralysis. The mental agony of recalling painful details was unnecessary and, therefore, buried memories were left buried. Tolerance for others is better than righteous anger and revenge. Assuming her problem was more than her father not paying enough attention to her need for importance, but a case of molestation, even that must be forgotten as life goes on. To have value and importance is the central issue of existence. To be something to oneself is the key. Assuring her unconscious to forgive and forget gave her a reason to disregard her anger as a mistake, not worth killing herself for, and to forget what was already past and look forward to the present. In fact, to be something of value to oneself is the key to explaining the purpose of the universe.

The idea to have value is already the "first cause" in the subatomic formation of bundles, which physics calls quanta, but to quantify by itself means more of the same. The point is that the bundles assume values and on the atomic level they materialize as qualities. The known atoms are more than electrons around a nucleus with specific numbers. They are hydrogen and oxygen. They are silicon and iron, lead and uranium, and, in fact, are the material values from which the physical world of values is organized. It is quality, not quantity, that has meaning.

Let us consider the second cause, organization. Everything is organized from atoms and molecules, but is organized by ideas that suggest value and meaning. If the organization begins at the subatomic level, then ideas organize the waves that appear as our reality, and ideas give the meaning of the identity of things. Ultimately, all that exists evolved from ideas that suggest value. Each entity that exists then depends on an idea of self-suggested value, suggested in the chain of evolution from a wave in the void, to a self-conscious person. This means that ideas only indirectly create worlds—that they excite the emotional psyche which generates the energy that results in the waves that physics has discovered. But because ideas are immaterial, physicists have a mental block. Reality is structured and limited by separations and we are committed to a physical existence while in a body that is separate in individual identity.

Many persons have a valid cause to feel discontented, particularly physically-handicapped individuals, but it is surprising to see that discontentment also happens among physically healthy persons, resulting from imagined feelings of injustices done to them. Such individuals are slipping into a negative trend, based on wrong expectations in their imagination that the world owes them effortless living or unearned fame. From anger against the world—a world one cannot change—the anger is turned against oneself. This causes most of the depressions so common in mental patients, affecting their physical functions also. Many persistent illnesses are indirectly self-suggested disorders. We already know from demonstration with hypnosis that suggestion can effect both body and mind. The question we should ask is, how can one prevent mental illness and recognize it early before it becomes a fixed idea?

In modern psychology, human nature has also turned out to have an emotional basis, which for a normal person is a positive disposition at birth, but which can become negative and self-destructive in the course of life. The problem is compounded by the fact that one must have acquired values in order to establish an autonomous identity. One must adopt values and assume a separate identity to have meaning for oneself. To bend backward to everything leaves one open to follow suggestions from others, and by giving in, will invite others to control us. Life is still based on the assertion of self-value and defense of one's own separate identity. Before reason and anything else can be applied, one must be a whole person with secured values.

It would be easy to assume that most of the problems in life are caused by fixed ideas. There is a healthy balance between positive and negative suggestions, and as long as one stays mainly on the positive side, that seems to be normal.

Without question, the young woman who was saved from certain death by self-poisoning had unrealistic expectations about human nature, whatever the intimate details. Perfect persons do not exist and justice can never be done in the world. One has to be contented with imperfectness and forget blame and revenge.

In life one must make choices, which are usually unfair. Falling in love is one example. By preferring one person one must reject others, because it is by judgment of value that one makes such a choice, which implies that the others were not as good. This is a personal judgment based on beliefs, but certainly is not an indication of real value. The is no absolute scale for worth. This is why one out of three marriages fail—because suggested value is based on imagined qualities. Values discriminate. Love, in particular, is always discriminating and unjust to all who are not included. Individuals in love will imagine that the loved person is the fulfillment of their expectations and will ignore all negative advice from others. Love makes one blind. Even a cool-minded, materialistic individual who thinks he makes his choice by reason, may be wrong. We seem to have latent images, unconscious ideas of beauty and worth that sway the choices. This again is where all understanding culminates, namely in the suggestion of value. The universe is a value-creating process and suggestibility is attracted to positive emotions on subconscious levels. Unprovable this may seem, because the effect is slow and is uncertain in detail, but the total picture confirms it.

Let us agree: In our real world nothing is perfect—the world obviously is designed to remain imperfect and incomplete to continue as a challenge for eternity. Such is the explanation of the continuum infinitum which the universe appears to be. Just for the record, if infinity is incomprehensible, so is finity as well.

Just because I also recognize the presence of struggle in the world, I still distance myself from Heraclitus and Darwin, from Machiavelli and Nietzsche, and from everyone who believes that struggle and violence stand behind the creative momentum in nature. I disagree with this perspective and instead believe that the purpose of the phenomenal world is to experience interesting events, and the brutal interludes are intended as reminders to keep us from losing our judgement so we will appreciate health and the value of life. In the design of reality, there is a strange dynamic of necessary imperfection and incompleteness, disregard for individual existence, and for suffering as if only the process were the issue.

Let me switch back to the time when I found out that my earlier understanding of a common principle was incomplete. After I became aware that the world process was not what I had expected—namely a straight value-creating process, but a retarding and conserving process, where protecting and establishing of values and identity was of equal importance to allow a temporal presence in the flow of change—I was made aware that one can always learn. I had assumed that I already knew the principle of reality, and thought, if one knew the main principle, the details would fall into place. That is true, but details have value also. At least the addition was in line with the general principle, which proves the principle was right and the addition was actually leading to a clearer definition of manifest reality.

After a period of rest, or as I would call it, incubation, it became clear to me that brakes are necessary to prevent evolution from running away, when time is plentiful in eternity. It can be understood that the only criterion of existence is the experience of meaning—

that space and time are background, and evolved values are a commitment. My understanding of reality has only become clearer as I have matured in life from father to grandfather. The insight that change is most likely to happen when no other alternative is left has meaning. Because change happens in jumps, this implies that nature wants to remain as it is, and if it must change, the change should be worthwhile. The result is the establishing of distinct separation of values, identities and separate levels. Also, that without fixing experiences in memory, knowledge would be lost, and evolution would be inconceivable. It seems confirmed that acquired knowledge of countless generations can be stored in the genes, as ideas for organization. But how can one prove that fixing in the real world actually occurs?

Storing knowledge and fixing it is commonplace everywhere. All learning is stored in the memory banks of the brain, because attention to the outside world has to keep the mind open. Knowledge that is not immediately needed is stored away. We also consciously help our memory by storing knowledge in books and find other forms of recording useful information that the brain would consider unessential. Even useful skills are rendered automatic to free our attention. For example, while driving an automobile, typing or reading, we can think of other things besides what is in front of us.

Suggestibility is proven to exist on all levels and so is organization. That suggestibility can affect organization is necessary to understand self-creation, but we have only a limited influence on creation, mainly on our own lives. Apparently our collective suggestibility can influence, in some uncertain manner, the future development of mankind, at least indirectly. Genetics is exactly that, multiple experiences stored in the genes from many ancestors. The wish to preserve and fix their values and identity is normal for anyone, and every sane being clings to life as long as it can.

Old cultures and traditions loyally hold out against change until they die from lack of interest. Violent overthrow in a revolt is not as frequent or even as useful as one might expect, because the mind must be convinced. Force cannot convince people, it only drives ideas underground. But reason can change ideas. What will happen next is uncertain since the future is unknowable. Yet this is the reason why life is interesting. It is uncertain, and there is always tomorrow. Even death is only the physical termination of an infrastructure of consciousness, because consciousness itself is immaterial and is the eternal essence of the universe. It is in a process which organizes and disorganizes, in what Heraclitus called the "play of the aeon." By aeon, timelessness is implied.

From the perception of the presence of organization in the real world, the new model organized in my mind before I knew the details. When innate suggestibility became obvious to me from hypnotism, it explained how the process of organization in nature works. All other details have been added later, such as the psychological interpretation, the nuclear and cosmological side and the experimental confirmation of organization by Gestalt Psychology. All fell into place and confirmed that the principle is correct.

The experimental confirmation of both organization and suggestibility to value occurred later, when the school of Gestalt Psychology, experimenting with perception, indirectly proved it. Prominent among the discoverers were Max Wertheimer and Wolfgang Koehler, who observed that the human mind has a "compulsion to organize" separate fragments into meaningful concepts or images—a gestalt. These men, as scientists, did not jump to grand conclusions, but rather referred to it as the compulsion for organization as an innate trait of conscious organisms. They spoke of a "Primat der Gestalt" as a commitment to a rational explanation. One is organized in a certain gestalt—male, female, proportional and beautiful or just functional, which predetermines human life (this is a given primat by birth). Gestalt is a physical concept. But I see more than the physical gestalt. I see the primacy of ideas, which organizes everything, even the organic gestalt by meaningful design. By enlarging the compulsion of organization to innate suggestibility to value, which as consciousness organizes itself into challenging worlds of meaning, the underlying principle and purpose of the universe imply the creation of value and meaning.

In fact, suggestibility was misunderstood, since even in the psychology of the unconscious suggestion was ignored as a side effect with limited importance. This gave me a chance to expand on this ignored field and base the organization of worlds upon innate suggestibility. Because it was a correct insight, things confirmed it. Yet suggestibility reaches beyond what is presumed to be real, into the transcendental and amorphous spirituality of suggested value, and opens other doors as well—doors that had been closed before.

The feeling that one sometimes has, must be connected with the universal mind in transcendental experiences is one of the fields of interest to be explored. It is sometimes experienced in special dreams. Most dreams show a lack of organization and meaning. That may be because we enter into other people's experiences and participate in the affairs of the world at random. We may, however, also enter into our own affairs, and lead double lives. The interpretation of dreams is unclear because the language of dreams is archaic and symbolic. Even if psychiatrists have developed methods to interpret dreams, they are the dreams of frustrated individuals with confused identities. There are clear dreams of normal persons with organized content that make more sense in a philosophical perspective.

The world, as I understand it, is a highly organized phenomenon of waves that organize themselves into meaningful entities, including self-conscious beings like us. This process of identification and fixing is by ideas and has proliferated into the incomprehensible vast variety of separate beings that express the intention of the whole.

In a process of self-creation of value and meaning, the meaning is received from its own opposite—low is opposite to high, or night from day. Without night, day would be without meaning and would not exist. We would not know that we are alive without death telling us that life can be lost. Each value depends on the other for meaning. It is as if each value has two opposite ends, as a magnet has positive and negative ends we call poles. The polarization of values is positive and negative, the first separation in the creation of meaning.

The unconscious is still a mystery, and transcendence is probably a higher level of consciousness that human beings occasionally experience. I shall come to this later.

I should emphasize again that I consider both major concepts, scientific materialism and philosophical idealism, as outdated. A new model of reality based on innate suggestibility fits the overall facts, and as far as the traditional ideals of goodness and justice are concerned, these only apply to the social level on which we humans exist and belong to a similar dimension to which the conscious mind and reasons of higher evolved beings belong. For instance, the poet William Shakespeare wrote, "all the world's a stage," thereby implying we are actors and life is like interesting theater, in which we participate and have roles assigned that allow us freedom of expression. This is a widely held understanding, similar to the idea that "life is a dream," and not quite real. There must be some truth to it, because many writers also speak from their unconscious. If the way to understand the meaning of being is through the unconscious, the opinions of transcendental seers are equally important to understanding the principle of reality as reason is.

From such seers comes the insight that ultimately nothing exists but consciousness. If one imagines himself to be a conscious entity in the absolute void, then the wish to become something of meaning—anything at all—is more than nothing. This is a powerful suggestion to explain the desire to create worlds and to participate in the experience of life.

4

Suggestion and Organization

I REALIZE THAT MY GENERAL OUTLINE of the Universal Principle raises many questions that demand answers, the foremost concerning suggestibility. Hypnosis is only a minor stage of a much deeper, far-reaching suggestibility in all beings. As far as the influence on the organization of organic processes and material objects by humans is concerned, suggestibility is insignificant, but it has far-reaching consequences on the explanation of reality by philosophy. In order to achieve any kind of trance or for hypnosis to work, innate suggestibility is necessary. We have discovered that everything is suggestible, not to our will, but to itself. Others can only indirectly influence us, but without our later approval by the unconscious, it remains futile. It was the philosophical importance that suggestibility has on the organization of the world, on nature and the universe that caught my attention. I was intrigued by the possibility of explaining most of the mysterious phenomena that have until now confounded philosophy and science. If organization always starts on the wave level of what we call electrical energy, then suggestibility must have existed prior to the manifest world because organization is basic to the creation of worlds.

Hypnosis, meditation, dreaming and hallucination belong to a group of phenomena known as altered states of consciousness, all of which require suggestibility to exist and function. Many volumes of experimental data have been written to which one can turn for details about altered states. Several forms of altered states are known, but innate suggestibility is required to make sense of the meaning of different realities, which is more than altered states. We know that insects experience their existence in a wave reality because they have electronic eyes. Bees would belong to such a class, obviously with the indirect purpose to pollinate flowering plants. Bees live in organized family hives, and are all loyal to one queen. There is no separate identity. A single bee will perish on its own. Loyalty is a factor in understanding the coherence of organized nature—a role that gravity cannot fill, not even for stars.

One can assume that with large organized beings the cells have organized into a single compact entity with a single consciousness, instead of just hives. Most higher organized cell states are larger animals which are organized into separate roles by a division into opposite sexes. The purpose of the female is to reproduce the species. The males contribute by fertilizing the eggs and protecting the family unit. For animals, such roles may be fixed. Self-conscious humans have freedom of choice. We know the fixation is very strong in animals and despite generations of living with humans, animals still live in separate realities from us, often seemingly dumb. The only connection we have is emotional. It is the same for all of nature.

This is different, however, for humans. We are flexible to suggestions by example or word. We can control impulses based upon innate suggestions because we are conscious beings, but totally repressing our nature would leave us disconnected. The strongest suggestions are through personal experiences in early life. Persons born into poverty believe that material goods are very important in life and they tend to be materialistic as a result. Yet, the average person also sees that ideas are important and material needs are only basic—that life is more than physical existence. Therefore, the average person usually has a broader view of reality than does the materialistically programmed individual. It is also a fact that we humans can be impressed by socially accepted suggestions of value.

Education is one way that social values are suggested and tradition is another. The existence of religious extremes proves that there are many cases of fanatical persons with fixed ideas. Although education is less radical, it also has people with strong ideas that are closed to reason. Sometimes it appears that even a scientific education can create deeply ingrained skeptical thinking. This fact is certainly enforced by conferring degrees on the best conformists. It is obvious that this is a process that is counteractive to new ideas.

In hypnosis, the experience of reality is altered through suggestion, either by another person who controls the ideas to be suggested, or through self-hypnosis, which is more difficult. But hypnosis is interference from outside. Because most suggestion happens unconsciously or indirectly, normally we may not even be aware of when we are being influenced. Whether it has a lasting influence depends on our inner identification with our values. It does not depend on whether the suggestion is true, but whether the unconscious believes it. We all have a personal judgement of value that causes us to decide what is credible or forgettable. This means passing experiences and thoughts can only indirectly influence our live. The larger unconscious must still approve them for value relevant to itself.

Suggestion is possible in the semiconscious state of trance that exists just before falling into sleep and when awakening. This interim semiconscious state, called somnambulism, also occurs in sleepwalking. Such words as somnambulism and hypnagogic state help to identify a meaning, thereby fixing it. This semiconscious state has deeper, far-reaching consequences. In fact, during this state the organization of many things can be influenced, because one is open to suggestion when the unconscious is unprotected. Philosophers have known about this condition for a long time. In the ancient Chinese book called *Tao te Ching*, Lao Tzu said, "If you think you know, you do not; but if you do not think, you know." This I take to mean that unconsciously we know, and call it "indirect knowing." Yet in a semiconscious state, we seem to have access to unconscious knowledge. This would imply that in this semiconscious state one is open to telepathic suggestion and transcendence is also possible. But it mostly occurs involuntarily, when it wants, meaning it is selective.

This limited response has also proved to be a barrier to actual mind control in hypnosis. There are limits, in particular, to the extent that the organization of the body can be interfered with. Such barriers also exist against all influences on the organization of anything in nature, whether organic or inorganic. That is why suggestion is bound to fail where physical changes are concerned and why experiments have failed to confirm that suggestion has influence on organization. But from this limitation, it would be going too far to conclude that there is no possibility at all that suggestion influences organization in nature.

We know that nature has evolved slowly and has organized in meaningful ways. We also know that nature is careful and protects itself against rushing to judgement based on a minor incident. Even in a hypnotic trance in the upper level, the unconscious can reject unacceptable suggestions that do not fit in to an already established design and identity. One cannot know whether a person has identified with moral values and acquired a conscience and therefore would resist unethical orders given by a hypnotist who is without ethical conscience.

The absence of moral values is an explanation of why criminality is on the increase in our present time. In our materialistic civilization, ethical thinking is associated with sentimentality and considered by some to be impractical in business. The majority of people obviously have low ethical standards since opportunism and competition is the basis of a self-regulating free society—which is better than controls or regimentation. It is certainly normal and natural that the strong set the rules, but we are civilized beings and do not have to follow nature in everything. Even nature is defensive against exploitation, including that of unethical hypnotists, and protects us by limits, because if the suggestion is contrary to our values, we suddenly awake.

Hypnotic trance can be compared to dreaming, because it is in dreaming or even daydreaming that we connect to the unconscious. In such dreamlike states, critical judgement is suspended and the imagination is free to roam. This state leaves a subject open to influence by what would be normally incredible suggestions, which are subjected to later judgement for value. It should have been clear from the beginning that hypnosis is only an aid to suggestion and that suggestibility is the issue. Even induction into a trance is achieved by suggesting tiredness. I expect that some people find that hypnosis is limited and does not work for them.

When I discovered that hypnosis was empty, I had not considered that the unconscious would later sort out irrelevant suggestions out and discard them. Only when values make themselves known is the influence of unethical suggestion blocked. Values are important and the unconscious decides whether something has value or is nonsense. Yet the idea of innate suggestibility and the belief that organization is affected by suggestion were confirmed for me. But humans have little direct influence on the organization, which is mostly decided on the unconscious level.

This understanding of purpose and motive in nature, based on identification with values makes innate suggestibility rational. It is not necessarily what we would call reason, but what has value for itself. Because of the limited usefulness of hypnosis and its short-lived effect most people think it is only a secondary side effect—probably similar to dreams, imaginary and unreal. I believe I have already transcended that concept. Hypnosis is an induction to allow suggestion to be accepted temporarily, otherwise it is suggestibility that is the real issue, which can explain reality.

The apparent complexity of the transformation of material objects into other forms and substances should not amaze us because atoms are suggestible on an exceptional level, but only to a limited number of emotional people who synchronize with materials.

Experiencing instant transformations of physical objects can be a shock to observers. However, the knowledge of electroquantum theory—that the universe is a wave phenomenon and the understanding of innate suggestibility—can take the mystery out of it. The transmutation of material objects in something else, even instant transportation to other locations, seems bizarre and illogical, but can be explained. What is missing in the scientific model is the explanation of meaning, because all things are organized with unique qualities and meanings of their own that give them value. Recent reports of psychokinesis indicate that sometimes one can see transformations happen, and that filmed or videotaped recordings exist showing transformations slowly taking place in the open in a couple of minutes. Where did this happen? In Brazil again and in China.[5] Another person who transformed material objects into different substances and gestalt has been investigated. Occasionally there was an unusual side effect that accompanied the transformation of materials, which had not been previously observed in psychokinesis, namely high energy electrical discharges. This particular effect gave physicists new hope that the universe is strictly electromagnetic energy. But there is still the issue of organization into meaning, which electrical energy displays but science fails to accept.

If one prefers to call the emotional energy electromagnetic, that is an option to avoid emotions and consciousness. But what about meaning? I prefer to call electricity also emotional energy. And in the case of meaningful transformations organized by a suggested idea, this confirms suggestibility in materials.

I am referring to the medium, Thomaz Morais, who lived in the region of Minas Gerais where the transformations were recorded about 1980 by many people.[6] It is interesting that these electrical discharges and lights were affected and inspired by Morais, a highly emotional and suggestible human, who claims to have been struck by lightning as a youth. The Chinese psychic Zhang Baosheng performed in front of video cameras in Beijing in 1990. Since then, all his work has been kept a secret. At Duke University in Durham, North Carolina, U.S.A., experiments with electronic equipment showed the influence of radiation by psychics. To many people, lightning and thunder are emotional expressions of nature. The fact that electricity is a natural force that discharges to the nearest conductor does not mean that it is mechanical.

This confirms again that so-called electromagnetic energy can also be explained as emotional energy. In a time when visual confirmation was impossible because film had not been invented, the work of African medicine men produced similar phenomena. In fact, ancient people have known about such events. The alchemists believed in them until scientific skepticism officially won, and the mystics were silenced or expelled.

In scientifically inspired Europe, the mystics and alchemists were driven underground, but in Asia the fakirs and yogis still remained respected persons. For all practical purposes, in normal life a few exceptions from the physical norm mean little—life cannot be based on rare exceptions. Practical persons developed useful and dependable methods by experimenting with material and energy, and eventually the majority won—materialism was believed. When physicists split the atom, they continued to split the fragments of the atom and called it "particle physics." Instead of less, they discovered a hundred or more subparticles to which they gave such names as masons and quarks. They had arrived at phantoms of nothing, because the particles are without substance. The explanation of so many more new particles may just be the result of the reorganization of waves into something new and slightly different. In the ground of being, the ultimate compulsion seems to be to organize into something. Anything is more than nothing.

Atomic energy

In the attempt to analyze nature, scientists discovered decaying atoms. Because these atoms radiated energy and purified in concentrated mass, scientists believed that atomic energy could become a source of power to drive mechanical machinery. The radiation, however, could not be harnessed and most scientists gave up. In Chicago in 1943, however, at the Manhattan Project scientists enclosed this energy in pure carbon and directed the radiation inward on itself. This was the beginning of the nuclear age.

What actually happened, in my understanding from the psychological side of materials, was that purified disillusioned isotopes radiating energy were turned on themselves and became a critical mass of emotional energy that approached meltdown. Luckily, the meltdown was halted when the increasing pressure from the radioactivity was released through a valve.

This emotional effect was used in atom bombs and to produce electrical energy. With the capability of self-destruction in human hands, the time had come to understand the much less destructive principle of suggestibility in nature. Let me return to average persons and more common experiences, to people who are satisfied that mysterious events are rare exceptions, who prefer separation of identity and values, and who are glad that a chain reaction is difficult to achieve. Although infants are born unconscious of the world, they instinctively know they have to suck and cry for attention. As the exterior senses are discovered, the infant expands its outer horizon and gradually organizes its experiences into meaning. At first it is a childish world, learned from crude trials during the first four years. Thereafter the horizon gradually expands and is frequently adjusted until, at about ten years of age, independent thought develops. All this is only possible because innate suggestibility is basic to self-orientation and is the essence of the consciousness that a person is born with. Without it, we would not understand anything.

With the development of conscious orientation to the outer world, the mind learns to organize phenomena into meaning and vision becomes a means of human perception. Thereby man clearly separates himself from the perception of animals and most of nature. There are occasional throwbacks, however, to telepathic thought and other frequency ranges that remind us of our animal past and emotional energy if we are cornered without escape.

Under hypnotic trance, the subject can often do certain physical and mental acts that in a normal state of awareness would not be possible: When done in a trance we accept it without question. Normally we are skeptical of our abilities, and while this skepticism is necessary to protect us from countless misleading suggestions, skepticism is also the reason for underestimating our own capabilities. Because of early failures, which have impressed themselves upon our memories, sometimes we doubt our own capabilities. In any case, such preconceived ideas are bypassed in trance and we find we can do unbelievable things. Clearly a temporary suspension of judgment is involved, implying that in a trance the values and meaning of things can be deliberately altered and suggestions uncritically accepted, but only until judgment returns to the unconscious skepticism and refers back to its own experiences after the trance dissipates.

Hypnosis is a complex topic which stimulates disagreement in details, but at least in general we have a good understanding of it. Regardless of differences in detail, suggestibility is very real and should not be confused with hypnosis, because suggestibility is always there, consciously or unconsciously. Even if the personal experiences that influence our judgement are subjective and limit us, so do the so-called objective facts which are often more limiting than our experiences. They reduce us to the physical level and the mechanical mode as subhuman.

Suggestion is a very interesting subject which would require volumes to do it justice, and I can only mention relevant aspects. Two hundred years ago magnetism and electricity were believed to be insignificant phenomena of nature. Now they are understood by physicists to be the ultimate sources of the energy moving the universe. But we already know that energy is selective and suggestible, so magnetism and electricity may have to be unified and also called suggestion.

Autism and fixed ideas

Just because a person wants to be hypnotized does not mean he can go into a trance. Unless the subject has a subconscious trust in the hypnotist, the process of hypnosis is difficult and time consuming, and may never lead to a deep trance. Some drugs can help relax this fear and achieve instant trance, but the use of drugs is a forced way to go into the unconscious and can cause problems with escape if values are challenged. With drugs the subject is exposed to any suggestion, particularly if "mind-expanding drugs" such as LSD-25 or peyote are ingested.[7] The high suggestibility which such psychedelic drugs leaves individuals open to dramatic experiences often on the edge of sanity, confusing their values. One becomes not just open to but unusually defenseless against the suggestions from single-minded objects that suddenly seem of grandiose importance. And what are these suggestions that are so important to these beings? Their own identities are fixed ideas of great value in the scheme of things. People should not try to challenge their ideas of value because it can lead to a dangerous confrontation. Their convictions of worth are unbending and have a similarity to fixed ideas in humans who are insane with grand delusions. Only admiration for this self-image can lead a person through this useless situation. Positive feelings can open doors to other dimensions of reality. Materials as well as people crave admiration.

Basically, matter consists of self-assured entities, autistic, withdrawn in to themselves, without sense perception other than emotions. We also know from psychokinesis that frustrated individuals can find sympathy from matter. In normal experience, matter seems dead, but we know it is pulsing with emotional energy. Physics has confirmed this part. The physical level is pulsing with energy and occasional outbursts of discharges do occur. Lightning and light flashes confirm this, as do earthquakes, which are emotional vibrations that transfer themselves in materials as waves.

The main fact is that the universe is emotional and not mechanical and has evolved to self-consciousness, a theory that physics has not yet accepted. Matter is on a different mode of being, unaffected by our values. Whatever we do does not seem to affect matter, unless we want to convert matter into energy by force.

In my understanding, the lack of interest in what other beings experience is intentional in the grand design. It is better not to know how others exist, how they stand being fixed to a static role. We humans are certainly on a wider, expanded range of consciousness than the so-called staple materials with fixed identities.

During regular meditation achieved without drugs a breakthrough is only possible when genuine admiration is projected, which causes the objects on which the suggestion of value is focused to open themselves and reveal their particular identities. It may cause them to return the admiration. This may remind us of the idea of love, but love actually needs an opposite to have meaning. This is why love is not universal but suggestion of value is.

Hypnosis is beneficial if used to free us from limiting, preconceived ideas, handicaps in normal life such as the unfounded fears based on unfortunate incidents in childhood. Yet hypnosis may often be unnecessary. Frequently people respond to rational explanations to understand the causes that may have been leading them to wrong conclusions. That theory—to analyze their wrong ideas—was Freud's preference, using psychoanalysis to find subconscious memories leading to self-understanding. This proves to be sufficient to free people from unfounded fears if they are still open to reason and not yet fixed and closed. But only admiration can open fixed ideas if suggestion of value has any chance which was the only way before drugs opened the closed doors of the mind.

For the purpose of explaining certain points of interest, here are a few examples. The subject of hypnotism is a special field, and health is not the issue of this book. I do not imply that I know details of individual cases, and I have no practical experience. But if the human psyche is a key to the understanding of the unconscious functions of the body, which is certainly suggestible, then I have a general idea. For special cases, hypnosis has advantages since it has the least side effects, compared to drugs.

Hypnosis is widely known in our time, mostly from use in medical practice by specialists believed to be trustworthy. For instance, when hypnosis is used to suspend pain during a tooth extraction or an amputation, it is particularly useful in stopping the blood flow. In a normal state, the body would not listen to a command to suspend pain or halt bleeding, but in trance it does obey. Actually, only a light trance is sufficient to suspend pain. Yet, after the trance has worn off pain is meant to remind us that an injury has occurred. Unfortunately, very few medical doctors and even fewer dentists can be found who use hypnosis.

Some individuals can suspend pain by telling themselves to ignore it. Autosuggestion works. In fact, it is the best method, but difficult to do. A common experiment in hypnosis is catalepsy, a peculiar state of muscle contraction that permits a subject to maintain abnormal postures without fatigue. This is proof that all the cells of the human body are suggestible, because the muscles consist of cells and all of them simultaneously obey the suggestion to become rigid like a wooden board. But the suggestion is actually selective. The heart is a muscle as well and knows that it has to function to pump blood and so it cannot become rigid. This selective intelligence proves that subjects under hypnosis do not blindly accept all orders.

If humans feel happy, particularly if they are in love, they glow and radiate energy. Women attract men by their radiation of value. The suggestion of value comes from inside, not necessarily from exterior physical beauty. This is another reason to think that the sun may generate more energy if it feels valued, which proves that the physical sciences are limited to a mechanical universe. What this implies, is that in the largest context, organization is guided by innate suggestibility to value apart from the effect that suggestion has on human cells and the human body. We now have learned that many living beings reveal this fact of being valued even more clearly and generate energy and health.

For example, many organisms living in the ocean have the ability to change the organization of their cells and transform the shape of their bodies. In particular, cephalopods (squid and octopus) can change shape and color by deliberately causing their cells to reorganize instantly. All this implies consciousness and means that nature is suggestible through and through, but still must be protected by separations from unwanted influence. To me, the presence of protection from influence was important which is expressed in autism. Other speculators dealing with hypnosis on a daily basis may not have taken it to the philosophical conclusion. Although hypnosis has been used for a century to heal and entertain people, no further progress was made. But, because suggestion was new to me, the deduction was logical that a system based on suggestion must primarily protect itself from the confusion of unfitting suggestions and therefore it could not be easy to use.

Because protection against suggestion is built into nature everywhere, hypnosis must be limited and difficult or the world would be unbearable and insane. I understand the need for stability and sanity in man to believe in a physically real world. In fact, the process of organization has produced a solid, manifested world to suggest realism. Evolution suggested a believable world, but the inquisitive mind of man was intrigued by minor inconsistencies in the surface and structure of the world and began to suspect there is something more than meets the eye. Behind the physical is the unseen metaphysical.

Besides, life is impermanent. As much as one may wish to hold onto the physical world, it is a passing experience. As one gets older and becomes aware that all things are impermanent and that one cannot hold on to life, one becomes open to the idea of a phenomenon intended to suggest physical reality. Yet, not all is necessarily lost. If it is not the material world that has permanence, then the immaterial consciousness that is innate could be permanent and eternal, perhaps recurring in a different organization. From all indications this possibility is confirmed. Extrasensory phenomena are reported by many humans. Transformation is universal, but does not necessarily imply reincarnation—at least not directly. But if our values suggest where one goes, that may happen after death.

Most people have this inner feeling of the survival of consciousness. Animals may have it too, but we have no proof. We will not have to wait long to learn the truth about reincarnation or transformation. Life is short, and we will find out whether the many speculators who believed in the survival of spirit were correct. All we can do is live. Since worry is harmful to living, let us not worry about death. It is more positive to anticipate a future than finality.

European Depth Psychology had already influenced me before I began this book. I believe that the unconscious is influencing life indirectly to avoid interfering directly. We seem to exist in two worlds, the internal and external. In fact, during sleep we communicate with the internal world. Hypnosis has gained a glance into the inner world and so do the dreams we all have every night, regardless of whether we remember them or not.

It was Jean Charcot, one hundred years ago in Paris, who demonstrated that certain mental patients could become normal during hypnotic trance and talk rationally, even about themselves and their illness. But as soon as the trances ended they immediately returned to their former insanity and resumed facial and body distortions. One becomes acutely aware that certain emotional illnesses are self-suggested "fixed ideas" in the unconscious mind of the patient, and hypnotic trance can only temporarily lift the self-imposed suffering that dominates the patient's reality. It is as if the patient wishes the illness on himself in order to escape from an unbearable reality. Perhaps this applies to many physical illnesses as well. Some persons want to be sick. This means that one becomes, at least in part, the designer of the direction that personal life takes, but cannot control the outcome. This can happen to children who withdraw into a state of autism.

Chance also plays a role in life. We feel free to be in control of our lives, but indirectly many suggestions can influence us. If we are unaware of them, we think we are free. Normally, one is constantly bombarded by all sorts of suggestions from which we are protected by the shield of a stable identity. During our lives we are constantly trying to remain independent from other persons and their suggestions.

From hypnosis we know that the more deeply the unconscious is probed, the more levels are found to exist. In deeper levels, the individual identity becomes even more diffused and the collective consciousness seems to appear and assert itself. These are levels beyond our influence, to which access is difficult and unadvisable. Access to deeper levels is impossible without the help of psychedelic drugs. Because it depends on mental disposition, a person can have a bad trip and never regain sanity. Drugs can create hypersuggestibility, and then control of imagery is difficult.[8] Drugs can allow us to enter the suggested identity in a level of value of physical objects, and it allows them to project their values and reveal their own grandiose self-admiration, which the subject has to endure. This information can be found in literature describing experiments with psychedelic drugs. Naturally occurring transcendence is a different subject. When channels open by themselves, it is not a forced entry.

When reports of "psychic surgery" came out, I made the mental connection that it must be some form of friendly transcendence, not the frustrated form of what has been called psychokinesis of emotionally stressed individuals. We can partially demonstrate that this is likely. Fire-walking is one way, and must be a positive approach since hypnosis alone cannot lift deeply imprinted suggestions of danger. Rare experiments with psychic individuals have shown that such feats are possible. This implies that a different level of deeper consciousness is necessary to transcend such fundamental differences of a major mode and level. From such psychological insights into the inner consciousness of materials and humans, one could think that the way physicists approach atoms, namely smashing or forcing them together may lead nowhere. To succeed in breaking up what was already a negative isotope of uranium is easier than to grow and organize one. From human beings we know that organization is inspired by suggestion of value, that positive inspiration is supportive to growth and negative is destructive. Could it be that attempts at fusion failed because the two basic expectations are wrong? The first expectation is that atoms are organized by suggestion of value, not created by fusion in extreme heat, but have evolved like everything else in nature in a slow process. And the second is that only a friendly and positive approach would create a chance to assure cooperation. The fact that growth needs and does not give up energy is logical. The idea of fusing living beings is inconceivable. It has no parallel in nature but is only done through artificial synthesis in chemistry.

Only on the molecular level is synthesis possible, but synthesis also requires energy and that would be the opposite from what is expected. That nuclear fission produces energy is logical because it is reduction of the atom's energy to a lesser level. But if one wanted to change or transform atoms, such as making gold from lead as the alchemists wanted to do, then one would have to "persuade" atoms by suggestion to transform from within. The transformation would mean reorganizing as another quality. On the organic level, this happens by cell division from within the same entity. Or it occurs from the outside when an transplanted organ is rejected because physically forcing two separate beings together is unnatural. From psychokinesis we know that transformation is conceivable but requires special emotional excitement to reach synchronization. However, this type of experiment approaches magic and nature has set limits to protect itself.

On our human level, trying the friendly approach—creating positive emotions—has been more successful. That is the way that nature is approachable. Ironically, that is what the alchemists once must have believed when they tried to make gold. They tried friendly persuasion to influence lead because they believed materials were conscious and influenceable, but it did not work. The alchemists did not know what we since have learned, namely that everything is separated by barriers to protect it from outside interference, protecting the identity of the individual. That would imply atoms are individuals, and are the most indestructible individuals on the physical level. On the subatomic level another mode applies. I think it is emotional, but certainly not subject to suggestion by our commands. That would be a catastrophe.

I could not inadvertently avoid challenging physics, and could not avoid expressing my views on gravity and, lately, on the nature of light. If the universe is really an emotional process, then there cannot be exceptions. Either there is a single principle that brings everything together in a single denominator or this model is another failure.

It may seem confusing when I frequently change from physics to psychology, especially when I imply that physics is totally wrong. This could make heads spin. Yet, practical physics as it applies to technology, based on causality, is still correct. It is theoretical physics, based on mathematical explanations, that is incompatible with causality in the real world and with their belief in gravity is unreal. For practical living, the world is what our senses perceive it to be. It should be taken as it is intended—to experience interesting lives.

What in physics, is called electromagnetic energy, I call emotional energy: Excitement within the waves of atoms and molecules that can become heated up to become light. Electricity is capable of all that because it is actually emotional as was proven by psychokinesis. If strong electricity is sent through a thin wire, the atoms in the wire become excited and the wire glows with light. A psychological parallel occurs with humans. When under stress humans can also become excited and literally shake and become emotionally overheated.

Mental illness must be mainly self-suggestion, because outside suggestion during hypnotic trance can sometimes temporarily lift the illness. A case reported from Britain makes this point clear. A man, angry and unhappy about some ill-chosen words he had spoken, made a vow not to speak a word for ten years. He kept this promise for a short while, but then decided it was not worth it and wanted to speak again. But when he opened his mouth to speak, it was impossible. He sought the help of a regular physician who reported that the man's vocal cords were in order and the doctor then called upon a psychiatrist to take over. He hypnotized the speechless man. During hypnosis this man could speak freely and said that it was because he was "very angry with himself" that he vowed not to speak for ten years, something he now regretted. But as soon as the hypnotic trance wore off, the patient returned to his vow of silence. During this ten years, he was frequently hypnotized and only then could he talk, but nothing permanently could be done until the decade had passed and the self-suggested spell had run its course. For this occurrence there are parallels in posthypnotic suggestion, but not nearly as deep reaching as this case was.

There are similar cases of self-suggested disorders, such as psychological blindness, when a person hates to see what is happening around him. An example of this might be a spouse's extramarital love affair. Normally, anger alone would not have such an effect, and other resolutions occur that let the hurt person accept it. But during heightened emotional states, without the possibility of escape left, it can become real, and persons have actually become emotionally blind, even though examinations prove that nothing is physically wrong with the eyes.

More common, but less obvious, is the general dissatisfaction with life that can turn some individuals to self-hate. When a person cannot strike back at the world, he turns on himself. The average person can take a lot of abuse, but without escape options the frustration can accumulate to a critical mass and a breakdown may occur. In nuclear terms, it can either be a sudden explosion or a slow meltdown. In human experience, an explosion is known as a berserk outburst of temporary insanity. More frequent but less intense releases of anger would have been healthier. Often, self-hate results in withdrawal into the self, shutting off all communication. At this point reasoning is useless. This is when the mind invents other realities or adopts a fixed idea. Some persons even assume another important identity, thinking they are famous or royalty, or even God.

We understand all this. If only the body would listen to reason and respond. But the unconscious stays unconvinced and if it has chosen to close doors to the world, it does. Cutting communication is a normal response, a common defense against unchanging conditions. Closing doors to intolerable conditions is a negative response that usually is only temporary. A new effort can restore normality if a change in the environment occurs and the person understands his problem.

Perhaps evolution is also partly inspired by frustration. But only if change can suggest an alternative way to a better design can we expect progress. A self-suggested vision of a better future is the solution in the human level of experience.

Psychology is a complex field in which sensitivity to individualistic values can explain many actions and hidden reasons, and that is why it takes years to solve complex cases. In fact, individual psychology, as Jung called his method for understanding the reasons why people are different, is one of many ways of bringing persons back to reason.

From transcendence and psychokinesis alone one has a feeling that the material world is a wave phenomenon, organized by innate suggestibility to value. With abnormal psychology confirming the emotional nature of the world, this feeling becomes even stronger, implying that the mechanical mathematical view is just not correct. The universe could be explained as an emotional phenomenon, similar in principle to our psyche, regardless of how colossal the dimension may be. The mode on each level differs to protect the structural integrity of reality by separation.

Art and culture

Another revealing comparison comes from art. Some musical people let their emotions range and thereby create symphonies and tragic operas as inner expressions of their feelings. Works of art are greater than normal life. The performing arts dramatize life by creating conflicts. Certainly controversy in plays, from comedies to tragedies, make theatre interesting just as real conflicts make life interesting. Artists express what the psyche of nature is seeking to experience, namely interesting lives, by providing mental and emotional stimulation.

And yet we do not always have to agree with the dramatization and tragic outlook of art on life, because negative suggestions can also call themselves art. Can we remain sensitive and emotional without lamentation and self-accusation, and still remain human? This question is academic because we have always interfered with the emotional sides of our nature and used common sense and intelligence to improve our lives.

While it is true that one must have an identity to remain stable and have value in order to survive, one should also remain open to reason. It is a sign of intelligence to be able to laugh about one's self occasionally. Some persons, however, do not seem able to laugh at life's foibles, instead, taking everything very seriously. They cannot forgive themselves, and believe that one must be perfect or be a failure. If a person is down on himself and the world, to change this self-suggested negative belief takes more than re-suggesting value. By now psychiatry understands that deeper suggestions affect the emotional level, and knows that recovery must come from within the patient. The patient, by his own insight, can and must convince himself to restore trust and find worth in living. Incidentally, trust is connected to loyalty.

That a deeper organic level exists we already know, since the body can survive for as long as it is cared for even when the upper level of consciousness is lost in a coma. Freud discovered the existence of the unconscious and its effect on the organic functions of neurotic persons who are still open to reason and suggestion. But closed-minded persons are at the next stage—beyond reason—with true mental illnesses such as schizophrenia, mania, depression and other forms of mental derangement, in which hallucinations and fear torment their peace of mind.

Fear is a paralyzing emotion that also immobilizes reason. Somewhere in the minds of the mentally ill, there is the notion that their lives are without future, and they give up hope. Such persons actually exist in a timeless limbo, as changeless beings in a constant "now," but still conscious in a state of autism. Even though the mentally ill give up, change is possible. They live in a permanent state of being, which is a grey existence without meaning, but there is hope.

In this world, bombarded by suggestion from all sides, a strong sense of self-worth is necessary to be able to exist. It must be the foremost concern of all caring persons that help is available unless organic disfunction has set in, physically destroying the nervous system. Other people can add to the confusion by their suggestions. Such suggestions are not necessarily ill-intended and, in fact, may be meant to educate, but they can prevent individuals from acquiring a personal identity. In the final sense everyone lives to experience life for himself. Individual integrity is necessary to remain normal and important to oneself.

This also means that too much love is oppressive. A person needs recognition of worth, and has to be something by himself in order to love life and remain open to change. This may also have relevance to education, because individuals perceive things with different priorities and mature differently. Schools should allow for this difference.

On the material level, entities have little chance for change and seem to exist in a timeless state with a closed and fixed identity. This resembles autism in children who start life already withdrawn, perhaps from some emotional experience of their mothers during pregnancy. They are emotionally vulnerable and withdraw to a lesser level of existence because it is safer to close doors to the world.

Telepathy or recall?

Particularly interesting for me is the field of memory recall and age regression. The complication is, because the subject is highly suggestible during trance, he is probably creating a fantasy. About thirty years ago, when age regression to past lives was the latest fad, *The Bridie Murphy Story* was published. It was the story of a woman who claimed to recall having lived in Ireland in a previous life under the name of Bridie Murphy. Yet personal details did not check out. Reincarnation was implied, but historical records failed to confirm that such a woman ever lived even though the general description of life in the seventeenth century corresponded to what other writers who lived then had reported it was. How did this woman, who claimed not to have read about old Ireland, know so much? It was eventually discovered that during her trances, she was reading the thoughts of the hypnotist (who was reading old books about life in Ireland) and she subconsciously wanted to please him. Whatever the details may be, it was the hypnotist, who unconsciously helped her to imagine her life in Ireland, thinking it would help her find her past. That is because telepathy in hypnotic trance increases. Whether or not latent memories are recalled from past lives, the subject is highly suggestible and thought transference is a common experience.

Since then past life recall has become a popular interest. Curious believers would like to know who they were in previous lifetimes. There is a problem in that besides telepathy, the unconscious has connections to the collective unconscious of mankind. Carl Jung had reason to believe that beyond the personal a collective unconscious exists from which dreams take their detailed information. This means that during a trance the subject could be recalling the memories of any or several individuals, because identity is uncertain.

Because science wants absolute certainty or nothing, it discards everything that is questionable. I think then it should start discarding its own questionable theories first. I can live with a certain level of uncertainty, giving me a future, but I do not want to know the exact details of how my life will end. The general idea of physical reality is enough to know. It is clear that the details are constantly changing, making certainty illusory and objective knowledge unreal.

Thought transference between humans is common and probably occurs during sleep when we selectively participate in other persons' affairs and ideas. While this is likely true, the identities of these persons are only general and symbolic. In sleep, as in a hypnotic trance, one is open to suggestions. But the values and symbols have strictly human meaning. This implies that one does not enter into the material level or the consciousness of other species. There are many questions about the understanding of dreams. Records from ancient Egypt and Babylon show that humans have collected and interpreted dreams for a long time. These meanings have been revised by modern psychology in our present age. Time has improved the interpretation of dreams greatly.

Mystics have tried to connect dreams to a higher dimension, but since dreams can also be vulgar this makes the divine emotional mortals like ourselves. Sometimes dreams are high spirited, and sometimes ordinary. Sensitive, suggestible individuals have placed themselves in hypnotic trances or used drugs to become transcendental, and then attempted to predict the future, just as earlier generations have done through sorcerers and magicians, shamans and witches, or whatever name they were called. We now call such persons psychics. The fact remains we have had only limited success and little has changed in two thousand years. We know that their abilities are limited to certain fields and that we cannot depend on their predictions in detail. It is possible that dreams and spiritual visions are perhaps from a dimension where time, place and identity are irrelevant. But without clear definition, such information, strictly by meaning, is useless by itself.

To know individual details in advance can never be allowed, and the stories we hear are in retrospect. Obviously the unconscious is protective of reality. Kings and presidents have used advisers, among them psychics and astrologers, but they did not help much, since they had only general information. The ancient story of King Croesus of Lydia, is well known. He had asked the Oracle at Delphi if he could win a war. The oracle predicted, "One empire will be won and one lost." He took this as an affirmative to his question and invaded, finding out that it was his own empire that was lost.

In a general way, uncertain precognition of events is possible, meaning that one is being prepared for something emotional to happen, but the information is not clear enough to do something to prevent it. Otherwise life would never be interesting.

For example, although we know that certain illnesses are terminal, we never know exactly when one will die. Since we are not interested in knowing such unpleasant details, we prefer not to think about death. And that is a critical point: We do not want to know details—only science does. Obviously, it is wise to be satisfied with general ideas of a meaningful world and leave it at that. The objectivity that science actually would prefer for calculations is not present in real nature since conscious entities can change, making certainty illusive.

Transcendence

From previous examples of hypnotic trance, it is apparent that trances can be extended to deeper and deeper levels and states, in which most preconceived ideas, even manifested objects of reality, temporarily fall away. Even the force of gravity which most physicists think holds the world together has been challenged. Reports are circulating that for short moments physical levitation has been achieved by yogis in India. If true, this would imply that gravity is also a suggestion of value, but mere humans only can achieve a minor effect of a few centimeters of levitation.

When hypnosis was discovered in Europe, it was considered a natural phenomenon, free of the mysticism of the eastern religions, just a method to suggest ideas to the unconscious and a part of science. Universal consciousness may in fact connect us all. But we have become separated in time and space by differences in value, and have grown so far apart that it may seem we could never be of one common source and follow one single intent. But evolution implies a common origin, and evolution is difficult to deny.

Incidentally, if the claim that physical levitation has been achieved is true, this would mean that gravity is also a suggestion. Gravity's purpose therefore is to keep things from falling off the planet, because it responded to suggestion. In turn, it would mean that the planets and the stars are kept in their orbits by the suggestion of value to be something important, not just mechanical gravity. The purpose of the world is to suggest a meaning for being, and thereby provide a motive for the organization of waves into meaningful worlds, which happens by suggestion of value.

Many humans have exposed themselves to extreme conditions of deprivation to understand reality. Among them were men like Gautama Buddha, and countless others, even many contemporary scientists who have sacrificed their health in experiments. One was an American, John Lilly, who described such experiments in his book, *The Centre of the Cyclone*.[9] He once came close to the experiences of insanity, lacking only the element of fear, since he still knew who he was. He also experimented with hypnosis, and compared the stream of consciousness to multiple computer terminals of individual minds, which all lead together to one great mind. Innate suggestibility always underlies meaningful experiences, and in our time computers suggest meaning which is relative.

The awareness of suggestion began with Franz Anton Mesmer, a medical doctor who believed that the magnetism in metals was also present in all living beings. He called this natural effect "animal magnetism." It was during the early studies of electricity and magnetism that the existence of electricity in frogs' legs was believed to be connected with the life spirit. Mesmer believed this and assumed that electrical phenomena affect human health. When he first used magnets on people in his medical practice in Vienna, he found that their ailments were relieved, but he was asked to "stop this nonsense" by the office of the medical profession of Austria.

Yet Mesmer was deeply convinced he had discovered something very important, and left for Paris in 1778 to introduce the world to animal magnetism by demonstrating its effect on human health. He opened a medical practice, and as his success with magnetism spread and his clientele grew, he soon found it unnecessary to use magnets, since people were cured just by his talking to them. Mesmer also set up a water fountain, from which people could drink and became healthy immediately. It became clear that he only had to be there in person for patients to become well. As a result, he was suspected of deception. Although he was a medical doctor and insisted that his discovery was innate magnetism, the medical profession did not believe in such nonsense. We understand now that he had discovered the presence of innate suggestibility, but the idea was before its time and also out of context with the mechanical models of physical reality. After watching Mesmer's success for awhile, the French Academy of Sciences conducted an official inquiry, from which they concluded that he exercised deception, causing people to imagine themselves being healed for a short time before their illnesses returned. The investigation was certainly fair, but Mesmer lost the legal case. The press denounced him as a charlatan who "mesmerized" people to get rich. Fearing for his safety, he fled Paris and vanished into anonymity.

The idea of animal magnetism was not forgotten by Mesmer's pupil, Count Maxime de Puysegur, who followed the idea further. He is credited with the discovery of hypnotism as a "mental function," which he called somnambulism because the trance resembles sleep. The word "hypnosis" came into the English language courtesy of James Braid, a physician practising in Manchester, England. Now we know what Mesmer had discovered was innate suggestibility, but this was unthinkable in a mechanical age. Regardless, suggestibility eventually became a logical conclusion. My observation that the organization of cells responds to suggestion occurred to me independently, not from a medical background but from a philosophical perspective to explain organization by ideas.

Suggestion is not just limited to verbal command. The body may not understand words, but it understands actions and feelings. Doing things also suggests value. Even seeing actions and visual participation has deep effects, while dreaming, much like acting in theater. Indeed, while dreaming we are playing roles which may deeply affect our subconscious. Unfortunately, from the point of content, the theatrical arts have more influence on us than we may imagine and like to admit to ourselves. The unrelenting bombardment of the public through the media by news and entertainment indirectly influences and confuses the values of people, implying meaningless and chaotic processes. And who is to say what is true? The only thing left to humans is reason.

One can forget about the prestige of science. It is less realistic than common sense. The universe is emotional and subjective and suggestibility is innate to all phenomena. This point was made clear by actual experiments in deep hypnosis without the fanfare of publicity. The story has been told that two good friends, both experienced hypnotists, wanted to try to go as deep into the unconscious as possible by mutually suggesting to each other to go progressively into ever-deepening hypnotic trances. But to achieve this, they also had to maintain a sort of lucid self-awareness. As they entered into deeper levels never reached before, they experienced a merging of minds and began to know the thoughts of the other. Then they found that they actually had contempt for each other. They had been friends before, but since each now knew what the other thought, they disliked each other.

Still the experiment continued until eventually they reached a point when they became concerned that they were losing their separate identities, especially when each had the sensation that his physical body was breaking away underneath him.[10] This experience made them feel like bodiless souls that their minds were entering a strange place, like a haven for bodiless beings. This dramatic experience was the final reason to end the experiment. Obviously only death was next, and these the two hypnotists were just not ready for that experience. What stopped them was self-preservation: They were not prepared to give up their individual identities and merge into undifferentiated consciousness, perhaps as fragments of the collective essence of mankind. Obviously, hypnosis can be pushed to the point where it becomes dangerous to mental and physical health. Yoga and Zen masters have described similar experiences during meditation, and have actually transcended the collective unconscious, and reached the last level where absolute emptiness broke upon them as the ultimate sobering experience of reality. Nothingness is described as a peculiar feeling of existing, but also of being free of any concern—as being nothing in a deep void, which I think must be boring and relaxing at the same time.

We have some understanding of meditation through the practices of people such as Raya Yoga. It is the idea of total self-control at the cost of giving up the experience of what life has to offer. This is the price of achieving control over the body and the rambling mind. One has to spend hours concentrating on directing his attention to known nerve centers to silence their activity and bring peace to the emotions in the mind. In this way, bodily functions and the mind can be controlled. In order to sustain this kind of control he must practically become a recluse, thereby missing the experience of life in an interesting world, which seems to be the purpose of existence. And for what reason? Is it to avoid living in the world created to have meaning, to avoid the experience of realism of the value principle, which is created by the existence of opposites? Is it just to avoid some suffering that is necessary for life to feel real? But then, would one also forgo the pleasures of life because it is true that without pain, pleasure would be unknown, and without evil, goodness would have no meaning? Opposites create meaning and value. From such ideas, it becomes conceivable that nature is creating interesting worlds from opposites.

It took nature aeons and countless transformations to evolve interesting worlds, to organize and manifest complex entities like the human body. Therefore, Hindu and Buddhist ideas to get out of life are negative in principle, although in many details, these philosophies contain rich experiences. I am certain that average individuals with common sense do not seek liberation from the cycle of rebirth to go back into nothingness, but look forward to a future. We continue to hope that we will reduce the suffering and as refined, civilized beings, to evolve to a higher level of self-awareness.

In fact, life is progressing upwards and seeks to rise to the next higher level. Evolution is slow, but while we evolve further we experience life. So there is no hurry. The experience of going somewhere is perhaps more exciting than the arriving. The future is unknowable, and exact knowledge would render the experience of life a forgone conclusion making the value-creating process and evolution meaningless.

To understand the intent behind the existence of worlds, one has to look at the meaningful ideas expressed in the world. To deny the presence of intelligence, for a mechanical explanation without meaning, is an insult to human intelligence. Even trying for a compromise, saying that consciousness is only a recent development out of matter, is missing the point. There is only one single principle.

What scenario can scientific materialism project but finality—that our world will freeze over before it burns up? Scientific materialism predicts that each individual will reach oblivion in nothingness. If nature is a process of differentiation, evolution is progressing not regressing. We are looking to the future, and this is the direction we should be interested in. Agreed, the future is unknowable, but if it is a continuation of the past, we should continue to evolve. And if consciousness is immaterial, then it is unaffected by whatever should happen on the physical level. One way or another consciousness shall continue.

From the existence of levels in the stratification of nature, we could expect that other levels will still be created, if they are not already in existence, with a different mode from our world—levels with expanded consciousness, where the experience of being is even more interesting. This is a positive expectation of the future, with the continuity of consciousness presumed.

All this is wishful thinking, convinced materialists will say. And on what is their thinking based? On mechanical concepts of reality, of course. To believe this, one must ignore the real world and believe that abstract science is real. If this is not wishful thinking, what is real? Let us take a glance at real life. I once worked with a man who was my equal in mechanical skill, and together we sometimes built tools. He had a habit of making sarcastic jokes at which we both laughed. It all seemed just in fun. He had recently been married, and his wife was a beauty. Later we had different assignments and over the years I saw him change. His sarcasm and skepticism that had left me unaffected became expressed in his face. He also told me that his wife had developed problems with her health, that his son was not doing well in school and his daughter had run off to live with a drifter. One day I happened to see him with his wife. Her beauty was gone, replaced by a bitter facial expression that to me seemed more like the distortions on the faces of mental patients. I saw him not long after this, and he mentioned that his wife had had a mental breakdown and he was also in therapy. In my opinion it was his sarcasm and skepticism that had become a habit which negatively influenced his life and his entire family. This situation took more than thirty years to manifest itself. My opinion about the skeptical attitude on which materialism is based is that it leads in a negative, destructive direction. This is a psychological interpretation of reality, in which suggestion influences organization, positive or negative.

5

Suggestibility in Evolution

EVOLUTION IS ONE OF SEVERAL IDEAS explaining how the world came into being. The oldest idea was that nature always is and has never changed. Another theory is that an eternal God is the single cause of creation. In Europe during the Middle Ages, monotheism replaced ancient polytheism. More recently, agnosticism and atheistic materialism have appeared. The rise of atheistic materialism can be connected to a scientific view that considers all previous assumption to be speculation. Science holds no beliefs other than experimental proof of physically real objects that are tangible and consistent.

So it was for some time, until Charles Darwin published his book, *Origin of the Species*. In this book he presented physical evidence of evolution which cannot be refuted—at least not on the organic level of nature. Organic evolution was accepted in principle by science, on the evidence of bones found in the ground. But the material level was still not explained, because materials such as stone are inorganic and seemed dead. One cannot actually know what happened millions of years ago, or even whether the word evolution is the correct choice for organic development in nature. This is still another question raised by philosophy.

Where proof is impossible, speculative reason can fill the gaps and one can still assume that there is a reason and a meaning to the world. Darwin and followers, as true skeptical scientists, were living at the beginning of industrialization in Britain, when scientific materialism was inspiring experimentation and when Newton's mechanical universe was based on the physical laws of nature. Philosophical idealism was considered unprovable. Pragmatic thinking became the order of the day. Darwin could not have explained evolution in any other way but as a mechanical and accidental process. He certainly could not have assumed conscious orientation at that time, without being accused of speculation.

One important factor today, which makes the observation of nature at its very source much clearer, is the use of video cameras in underwater studies. Here evolution as transformation is more visible. It is in the ocean, the cradle of organic life, where we still can see much of how life evolved in stages. What we now can observe is the variety of ideas which are trying to organize into something different and unique in a vast kaleidoscope of life-styles. There are already different ideas, from simple wave-like movement to wings implying flight, from crawling to walking, that were tried out on the bottom of the ocean. We can see heavy armor and even the use of jet propulsion. That ideas were experimentally tried out points to intelligence, to alternative choices, to evasive actions, protective design and cunning. We have the overall impression that there is meaning involved in the evolution of nature. In fact, many of these beings in the ocean are suggestible and telepathically receptive to our thoughts. Orcas, dolphins and octopus are intelligent and can anticipate our intentions. Others are at least conscious and avoid capture.

Particularly amazing is the organization of the physical form and function—the gestalt. It is so vastly diverse and meaningful in different species that higher intelligence and orientation to meaning must be involved. Accidental evolution is inconceivable. Although people think their ideas are unique—artistic self-expression, beauty in design, color and grace—in fact all are found in nature. Nature is not just utilitarian, it is also artistic.

One has the impression that over two millennia ago Socrates and Plato had concluded correctly that ideas are expressed in the creation of nature. But they were mistaken in assuming that the ideas are immutable. We know that ideas have evolved, that everything changes in time, and that ideas suggest the organization of the diverse beings in nature. If necessity and utility were the only basic motives, nature would be uninspiring. Perhaps there was such an early stage during the age of the reptiles. Yet, the variety of species indicates the undeniable wish to be different. This means that the unconscious intent to be something special is already there, and needing to eat, for example, plays only a secondary role. Darwin or any other materialist who assumed that evolution was achieved from the selective effect of the survival of the fittest was mistaken. Besides, before one can survive, one must already exist.

It is true that necessity can stun and exterminate life. But ideas are positive suggestions of meaning which survive despite many obstacles. Nothing is created from negative ideas except stunted growth and disorganization. As far the control over the organic function of transformation by suggestion is concerned, there are many example of nonhuman organisms exceeding human limits to transform themselves.

The octopus, for example, has the ability to transform, starting with a bizarre design. Its stomach is next to its brain and eyes, and its eight arms trail behind it when it swims by jet propulsion. With the stomach sac in front, it deceives its prey by appearing harmless, but as soon as it gets near its victim, it swings around at lightning speed. Then, using its arms to hold the victim, it eats it. But most interesting is that the octopus can transform its physical shape in many ways, looking totally different in each transformation. It can even change colors and the texture of its skin. If pursued, it releases black ink to blind its enemies, and is able to squeeze itself into narrow crevices where most cannot follow. This is a creature which has control over the arrangement of its cells at instant command.

Organization by suggestion is a critical point. It means that suggestion of ideas controls the gestalt in which to exist at will on this level. By sorting out what is possible and rational, one can arrive at a theory of how the world came into existence, namely through the evolution of ideas on how to organize in meaningful ways in order to experience life. In this case, the first cause is the existence of some form of consciousness that is suggestible and interested enough to participate in the creation of worlds in order to experience interesting events. Such events must be created by meaningful interaction of diverse beings with different values.

Because the procreation of the human body and mind happens inside the womb and there is unconscious organization inside the womb which protects the embryo from outside harm, one cannot see how cells organize. Therefore organization seems less important to science. It also indicates that intelligence is involved, something that a mechanical explanation would rather prefer to ignore. This could mean that evolution occurs consciously by ideas. But we are unconscious of the inner organization that takes place. We do not know exactly what life is, but we know what it is not. It is not mechanical and does not evolve according to the survival of the fittest as theorized by Darwin or the dialectical principle of Georg Hegel. We only now have evidence that innate suggestibility is the nearest explanation possible.

Evolution is not confined to organic life, but begins in pre-material wave organization. From our present knowledge about the physical structure of materials, the material level confirms meaningful organization in atoms and molecules. The material level that we perceive begins with the organization of atoms, starting with the simple hydrogen atom and continues to about one hundred different combinations in atomic values. It then changes to a different mode, into the complex molecular combinations of atoms in countless chemical organizations. This change of mode is a transformation of process that continues to the organization of crystal structures as materials or of cells and organic designs, in which meaningful ideas are expressed. This process is called morphology. It is on this level that evolution becomes observable, because time frames in which a transformation can take place are much shorter.

When cells organize themselves into mobile beings, moving independently and searching for opportunities to advance themselves, conscious orientation is undeniable and certainly intentional, not accidental. Because mobility prevents beings from developing roots, they must feed on other organisms that have converted energy from the soil and light. This we understand as symbiotic dependency in a food chain. The general organization of nature obviously is guided by demand and supply, resembling our economy in principle. This guidance also can be explained by suggestibility. The organic beings know instinctively what is good for them. But just to eat is not enough. To be something different and of meaning is the central reason of organization in which utility is secondary.

Suggestion somehow must manifest ideas into physical substances. This we know from brain chemistry: The brain produces the chemicals it needs to regulate the body. I could go on to explain the organic body as meaningful organization, and would arrive at conscious beings that orient themselves to opportunity in their environment. But in the recent past, a totally new understanding has been gained by the introduction of the wave concept of reality which explains everything in a different light.

The electrodynamic wave concept of reality in a philosophical interpretation is based on the separation and distinction of waves and their organization into temporal manifestations. The organization of sound waves into pleasing harmony, into great concerts with emotional impact, is achieved by the combining of sound waves into meaningful melodies. The separateness of waves is the same in principle as the separation of values, or the separation of things and beings. This makes the organization of waves into atoms the beginning of physical evolution, and evolution is a value-creating process organized by ideas which suggest the impression of reality. Basically all things are a phenomenon of emotional wave energy, organized by suggested ideas. This pattern applies to all electronic devices, to organic and inorganic organization of value. All depends on the separation of waves and their organization by ideas.

In the case of electronic devices, our ideas are organized by the conscious minds of talented individuals, but not the organization of living beings and of nature. This happens from unconscious ideas, although we may contribute indirectly. We also know that the overall organization, although beyond our intelligence, is not without error. We know that imperfections in organization do occur. The question is whether they are caused by an emotional disturbance on our part or by errors occurring in nature. It follows that unconscious directions by genetically stored suggestions of organization are real. With such ideas of reality, evolution is mainly directed by unconscious ideas.

The evidence supporting evolution is from the remains of bones which have absorbed minerals and become calcified. Other bones and cells dissolved long ago. A few fragments of a living past, of which only the physical structure is left to explain the meaning, are inconclusive. Speculation and reason are necessary to organize a model that comes close to explaining what this world is all about. And if the explanation falls outside experimental proof, which it does, the physical sciences are inadequate to explain how evolution happened. It is unrealistic to have a model of the world that ignores consciousness. If one ignores consciousness, then one is only dealing with the empty infrastructure.

I am convinced that the world has organized by meaning into a diversity of separate levels and ideas with complex designs. But the meaning of the word evolution has been stretched. If it originally meant gradual change, this it was not. Sometimes, when a major change of mode occurs, it could be called revolution. In fact, it could also be called creation by suggestion of meaningful ideas. The only actual happening was that expansion of consciousness took place and organization evolved further. In this sense there was evolution, evident in improved organization in steps but not flow.

Evolution is therefore a self-creating process in meaningful steps. Most beings are guided by their sub- and unconscious instincts, which is the common way to exist in nature. Self-understanding to the level of clear thought is only reached in the human species, with the organization of a three-level brain. But most beings are endowed only with enough exterior awareness to survive within the meaning of their species, and the idea for their role in nature is above their intelligence.

It is still speculation whether we humans descended from primates or have evolved from a separate line. But in regard to intelligence, we are certainly on a higher level of self-consciousness than monkeys. Chemical tests can only test the chemical arrangement in cells. Without the suggestions to organize, the design of the gestalt, or the organization of the brain is inconceivable. We know from probing living humans that the ideas of organization might be stored in chemicals as memory and may carry the suggestion that organizes humans differently. Only recently did we start observing electrical activity that occurs during the thought process. This could mean that consciousness has electronic character, which causes brain waves and confirms the electrical nature of consciousness.

Charles Darwin and Alfred Wallace explained evolution as the result of natural selection. This explanation misses the point. To select the fittest from an already-existing species does not address the question of how the species itself came to exist in the first place. In fact, genetics is doing exactly this, selecting the best from already existing species, by breeding for desirable features that are variable within existing species. This process creates nothing new in structural design. It is only a variable of the idea of a species.

The first error made was to think of nature as a passive process, accidental and unintentional. It is, instead, an active and aggressive process, born from the cosmic fire of strong emotions, with willful intentions to create meaningful worlds. Whatever has and still is happening is certainly not without self-assertion. Life wants to experience meaning, and the moment a child is born it asserts itself by screaming and by willfulness.

The idea that the unconscious creates by suggestion can be experimentally proven by the effect that suggestion in hypnotic trance has on the organization of cells. Let us return to established knowledge for which experimental data is available.

We are the product of a long line of ancestors whose life experiences presumably are stored in some sort of memory banks, which can account for unconscious memories. These are passed on from life to life by the genes. The question is: Are suggestions contained *in* the chemicals that make up the genome, or is the *arrangement* of chemicals enough? Certainly experiences and memories do not show up in any chemicals from the ground, but may be imprints of suggestion. This same applies to the chemicals in the human brain that, when electrically stimulated, recall stored memories. I refer to Wilder Penfield's experiments in locating memories in open brain surgery. I think that chemistry may not yet be the key to inheritance. It could be electronically stored information imprinted by suggestions, which are of value to the organization.

Before Darwin, there was little consideration of evolution. Philosophers never believed that animals, trees, insects and all the transient things could have created themselves, but believed rather that they were created by something more intelligent. That creator was a wise being who had the entire harmony of the cosmos in mind—namely a divine creator, God. But because we learned that creation happened slowly, by trial and error, it seems more a self-creating process, still continuing and forever imperfect. The other logical conclusion is that the unconscious has intelligence. It is in all of us. It is behind the whole of nature in the metaphysical consciousness, and it is still learning as we do. But this intelligence is more careful when it creates and could be a collective mind that is present in all beings of nature.

Since ancient times, philosophers have thought that a rational principle with a purpose underlies all of nature. In Greece they called it the teleological principle. In India philosophers understood the world as a spiritual phenomenon or illusion. Later, European thinkers thought the world was created by a theory of the universal world soul known as panpsychism. This is the direction of idealism espoused by philosophers such as Hegel—to assume that the world spirit is creating everything. Idealism has persisted even when many other concepts have changed. It has also remained unaffected by many scientific theories, all of which have briefly announced themselves, but usually have left too many factors unexplained and had to withdraw. Science in general leaves consciousness and reason out of its calculations—it is more comfortable with materialism.

Genetics

When a problem cannot be resolved by discussion, it is usually relegated to further experimentation. This happened to evolution. Biology was already experimenting with the development of improved strains of animals and plants by breeding with predictable results. By crossing the general idea of a species, such as humanity, one can cause the reduction or enhancement of a particular feature, as long as it already exists within that species. And even then it will sort itself out again by original color and shape in a few generations, as Gregory Mendel discovered. But biologists have failed to crossbreed species that reflect different ideas of design.

As part of the natural evolution of species, genetics can be discounted. It is impossible to crossbreed different species, even if they have a similar physical design because it causes confusion of identity. This is a critical issue for scientists and philosophers because it casts doubt on the idea of gradual evolution through adaptation. Evolution happens in steps creating meaningful values, based on ideas that work and allows a separate organization to survive.

I want to say a few words regarding genetic engineering by experimenters who are trying to insert alien genes into the genomes of others, to cross different species. Obviously these genes carry confusing ideas of identity, and result in bizarre forms, usually able to live only one generation as a hybrid. So the experimenters still have not created anything new which can procreate itself.

It is true that we now use hybrids and this has improved yields, but evolution is a continuous process and evolves from within by the suggestion of value. The concept of change from mutations is already obsolete—only deformed individuals with degenerative diseases descend from damaged genes. In the genes are inherited the instructions from a long line of ancestors. This suggests that a fixed design laid down like a blueprint, in the organization of the organism, is most unlikely because that would be almost saying that genes are our fixed destiny. Even if the genes have control over physical development, our conscious mind has control over our impulses. We cannot allow ourselves to act unreasonable. In the civilization in which we presently live, among machines and traffic signals, we have to live by logic and must be flexible.

Studies done on a number of one-egg twins has convinced many that genetics affects the total personality because it determines our private choices—the type of wife, the hobbies, interests and so on. But this may have a different reason also. It so happens that one-egg twins have two things in common—the genes and the same astrological sign at birth. Subtle influences from heavenly bodies affect our emotions, and our emotional lives, but not the rational behavior where the influence of the genes is limited. There have been, however, mental connections confirmed between one-egg twins that point to similarly unconscious influence on each other in daily activity which could be explained by astrology.

In fact, we are not as controlled by heredity as we think we are. We can alter our genetic predispositions by adopting healthy life-styles just as we can ignore the influence of the subtle suggestions from astrological constellations. There are many other suggestions from different sources that bombard our consciousness. Many suggestions come from other humans—even from the collective unconscious in dreams—all of which are trying to influence us. A healthy person has a strong shield against unwanted influence, selecting what has value and meaning.

Let me make a reality check on where I stand compared to current thinking in biology. While completing this book I found a new book published in 1996 by the renowned physicist and biologist, Arne A. Wyller, called *The Planetary Mind*,[11] which I think represents current thinking in academic circles. Dr. Wyller agrees that ideas can explain the intelligent organization of nature and that somewhere in the planetary consciousness a higher intelligence must be assumed because evolution is meaningful. But he still toes the line of scientific materialism when it comes to details. He writes that Newton's assumption of the existence of gravity was a brilliant idea, although it can not be proven by physical means. Gravity is necessary in explaining the movement of planets and the formation of galaxies and allows us to assume the Big Bang theory. How else can one explain the creation of complex atoms which require extreme temperatures to synthesize? Wyller's book reflects the dichotomy between idealism and materialism that has reached into science.

I think that most physicists are behind this paradigm of gravity. This means I stand alone based on reason in thinking that slow organization of "waves" has created the complex atoms by the suggestion of value. But the direction one chooses to view the world process is critical. By reason, from the most highly evolved, the human beings on our planet, the common principle of reality should be revealed. We are emotional and suggestible and so is the energy which created the universe of which we are minor replicas in the organic mode—a single principle is the issue here.

If there is an explanation of evolution, in particular of complex organized entities which already have an idea of themselves, it is the idea of their identities—that must change to transform. Only minor functional changes are possible. The organism, in order to function in a different manner, has to be redesigned. Nature does not bother with elaborate additions or repairs of living beings. It replaces old, worn beings with young new ones.

The evolutionary transformations apparently are planned in advance. After working out the rearrangement of the bone structure, legs and feet between man and monkey, it is the function of the brain which must be greatly reorganized. There is no evidence in nature that gradual transformation occurred. Instead, ideas suggest themselves first, before major changes of great consequence in function are implemented. The evidence for meaningful design changes from ideas is overwhelming, and on a level where we can see it. One level is that of insects. We can observe functional transformations when grubs change into completely redesigned beetles or when soft caterpillars turn into structured butterflies. It is a total transformation of a design by a new viable idea without involving gradual changes. The question is where does the idea come from? It has accumulated as suggestion of value over a long time, finally reaching a point where it can decide to make a dependable and functional transformation.

We have a simpler parallel in the transformation of atoms, which remain stable until their electrical charge reaches a point of saturation and they jump to the next higher level. The same thing happens to the generation of ideas in the human mind. After saturating the mind with a problem, eventually a point in the unconscious is reached where a solution is suggested. It is the same typical delay—a hesitation to let go until a critical mass has accumulated—when an overdue new idea is suggested. This process repeats itself everywhere. The delayed step of evolution has a slight resemblance to the dialectical principle proposed by Hegel, except it is motivated by innate suggestibility to value, not by external forces from creative conflict.

This pattern of accumulation before a change is physically implemented, is repeated everywhere in nature. On the technical level, we have to accumulate knowledge and develop the ideas for new designs in our rational minds. I know that many who think of evolution as gradual change are searching for the missing links between monkey and man, but there may have been a radical transformation, a jump.

Negative ideas

Nature is imperfect. Occasionally mistakes happen when physically and mentally handicapped humans are born. Imperfections in organization appear to arise from confusing suggestions with no value for normal living.

Suddenly I saw what is happening. Suggestion can also have negative influence. One can become immersed in negative thoughts and create negative ideas—ideas that are destructive to the organization. While many illnesses are already known to be self-suggested, sometimes the effect continues in following generations. Certainly, a negative outlook on life is self-destructive, just as is the harboring of hate and anger. It either turns inside, to cause health problems, or it turns outward as aggression and violence.

My opinion of evolution is based on the massive repetition of the same pattern of separation into distinct values, of which the clear separation of species is one example. All humans are basically alike in design but they are individually autonomous. If a memory is assumed to be necessary for organization, this memory is transferred through the genes from life to life. But the general design could not be influenced by the experiences of every individual, considering the diversity of conditions in which humans live. It must rather happen by a general consensus of the needs of the majority of individuals. We are already a long-established species. New suggestions must be in line with the existing design and values, compatible with the already established identity of the species. If life is meant to be a meaningful experience, it has to allow enough time to establish an identity, a chance to grow up and experience meaning. This means that change must be slow, and stability must persist long enough to establish an identity for a while.

Thus evolution is a continuing challenge between wanting to speed up change and wanting to hold back to establish an identity with meaning. In practical terms, this means remaining alive as long as one can hold back death while inevitable change makes temporal manifestation as a physical being a practical fact. Once manifested into physical form, the manifested organization is unalterable until it dies out through disorganization.

As has been mentioned repeatedly, there is another side to life besides the physical process, namely the conscious experience of an interesting life. It became clear to me that science and technology are experimenting only with the physical level, because it is stable enough to be worth their efforts. Therefore they speak of objective and subjective knowledge when the question should be whether something is stable enough to suit them.

Small thinking is inadequate for big ideas, of which evolution and the purpose of the universe are the largest. My intention is not to degrade science. It is useful, but it is just not relevant to understanding the principle of reality and the meaning of being. Scientists have told us that life is meaningless in a material universe. This tells us that their opinions are negative and depressing. Nevertheless, I can use some of their experimental data in a positive way, because the principle of meaning also repeats itself on the physical level in mechanical designs. Indirectly, science and philosophy contribute to the same goal: Understanding the metaphysical principle of the phenomenon of reality.

From psychology and psychiatry we learn that to survive one must have a meaning for living, a purpose for wanting to exist in the face of constant obstacles. This requires a strong conviction of self-worth, which can take a lot of knocking. Self-orientation to value by innate suggestibility is necessary in order to provide a rational explanation for evolution organizing worlds with meaning. The physical part is merely the infrastructure for the evolution of consciousness. It is a frame of reference on which consciousness supports itself.

Technology and mechanization

If, in principle, evolution obviously repeats itself on all levels, then it also repeats itself on the practical level of current technology. Perhaps we can learn something about the secret of evolution which on the organic level is unconscious to us, but on other levels is more clearly observable.

A good example of organization in a large-scale, commercial manufacturing process is the auto industry. This industry exists not because we have the technology to produce automobiles, but because people think that owning a vehicle gives them a greater range of mobility. This is the real issue. Without a suggestion of value as reason, this industry would never have developed, and the moment a better way to achieve personal mobility is provided, the industry will become obsolete regardless how ingenious it is. Of course, a profit motive is involved, but without practical value for people, profit is elusive.

The organization and development of this industry is most interesting. The need for mobility was always there, but not only was horse-drawn transportation slow, it required horses, stables, barns and a coach to sit in. The automobile is basically a coach, which may be stored in a garage or not; it goes great distances without needing physical rest. The demand was there for a mass product, and gradually an industry evolved which required organization on a large scale. Because the automobile has a complex design in which the ideas of many inventors were combined to develop a reliable vehicle, the organization took many years to evolve. To meet the great demand, systems for mass production are developed from ideas they planned years in advance. Prototypes are made and tested, before the tools to mass produce are built, so that several years later reliable automobiles can be sold.

The average car owner is usually unaware of the planning and organization involved in producing his car. This is also the case with the unconscious organization of living beings. We inhabit a body about whose planning and organization we are mostly ignorant. But the human body must have been planned by some intelligence, if not a design department, because it is more complex than any machine. It requires conscious orientation that no mechanical device can claim to have.

Why this comparison? Because evolution happens in the creation of a complete and viable model. A human model change is not as likely to happen as quickly as change in a mechanical contraption. A major difference is that our body has organized itself from within by adding new completely viable levels of consciousness on top of previous levels. On the last layer of external consciousness we developed our external orientation as judgement of values (also called reason) which made us thinking, autonomous beings.

But machines fall short in regard to consciousness, probably because they are assembled from manufactured parts instead of growing naturally from within. The result is that they need humans to control and service them. In fact, without value to humans, machines or electrical devices are meaningless to themselves.

If in principle the same general pattern repeats itself everywhere without exception, whether natural or artificial, organization is also displayed openly in the manufacture of mechanical contraptions. But in order to exist, anything that can survive for a while must have value or it will not exist for long. Even self-creation must also have started from a suggestion of value as an idea. Nature must have planned and tested many prototypes before it created viable beings that function as long as meaning suggests itself. Nature is careful but occasionally mistakes in organization happen and are usually aborted. Still some failures in organization do survive controls can happen because nature is learning and continuously evolving. But individual suggestibility to value has as much influence on organization as a single buyer of an automobile has on the design and manufacture of automobiles—only marginal and indirect.

When change is made, it will be from a consensus of many experiences and will represent a clear step up—not merely a few minor improvements, but radical enough to make a difference from what came before. This has resulted in the evolution of distinct species or models, sometimes even complete transformation of the design and idea. The proverbial quantum leap is a parallel of the same pattern recurring on all levels, making changes only when necessary.

Separation of values and meaning is the obvious intent of evolution. Organization raises the level of consciousness and with it, a refinement in awareness occurs. On the biological level primitive life forms are dependant on each other in the food chain, cannibalism included. More highly evolved organisms, such as humans, have a greater self-awareness and sensitivity to pain and suffering that results in the recognition of possible consequences of their action. In the present state of civilization, this empathy has developed in mankind at least on individual levels and has created compassion for other beings.

It appears that the system of nature is designed in such a way that only worthwhile changes occur, sufficiently different to be able to make sense. This is another reason for the delaying pattern in evolutionary progress and it explains why changes occur in steps and jumps. This is how evolution progresses. Every step seems to level off onto a plateau before the next step is taken. Once personal identity is established, this identity may be carried along to a next level, but altered in mode by that level.

From our perspective, significant evolutionary changes are often delayed and seem long overdue. When they happen, they happen suddenly, as viable and meaningful transformation—distinct in value. Biological changes in a species would be accompanied by broad chain reactions which must transform many individuals at the same time. Less than one hundred would become outbred. More transformed individuals are better in guaranteeing survival. All this is a logical conclusion from the real existence of distinct levels that separate everything in nature. There is nothing that resembles a flow of change, but wide gaps between species and modes of being, stratification of distinct differences in value.

One major step in the evolution of our planet was the creation of the biosphere, a layer of organic life found nowhere else within the solar system. Unless we are psychologically blind to other levels, we must presume we live on a unique planet. This presumption is suggested by the special reality in which we exist. But the mind can transcend the limits of perception and think it possible that other planets with a biosphere could exist in the vastness of the universe or even on the barren waste in our system on other frequencies. In nature, psychological blindness seems to be the rule, existing as a form of autism towards more highly evolved beings. In fact, cases are known where humans became selectively blind to certain unpleasant facts that could have undermined their self-value. I am not saying that we are psychologically blind, only that we could be to levels near to us that we subconsciously prefer to ignore because they would make us feel inferior.

We humans are suggestible to opportunities and new ideas that improve our lives But we are reluctant to change our personal values, and do not easily trust single experiences. We need many repeated experiences to dramatically affect our lives and influence them sufficiently to change ingrained ideas. This explains why shock therapy works, and why tragic experiences have deep emotional effects. This also explains why people often make irrational and illogical decisions: They trust the logic and advice of other humans less than their own experiences.

Metamorphosis

We humans may never know how many different species there are in existence on our planet, or how they evolved in detail. But by comparing other species to ourselves, we can have a general understanding of what may have inspired them to evolve. On the insect level, where life is short, we have a glimpse of how change occurs in nature. One of the many examples is the well-known metamorphosis of the caterpillar into the butterfly. From the outside, we can observe how relatively quickly this transformation takes place. The advantage of change to be a caterpillar is obvious. As a caterpillar, it moved by crawling on trees to eat leaves, obviously it would be much easier to fly to other trees. This idea has value and may suggest itself to even lower levels of consciousness. While the cause for change may also be that flying could be an advantage to ensuring propagation of the species, this is also a suggestion of value, and both ideas together could be the true reason. Certainly the stage between—half-crawling or half-flying—is impractical.

In fact, organisms in nature are designed strictly with expedience and economy in mind. Nature is superior to any of the mechanical devices that science and technology have invented, which are all incapable of change. The metamorphosis of the caterpillar, taken to the cell level, is a major transformation of design, representing a total reorganization of the cell structure in a very short time span. The caterpillar knows little of what goes on inside itself. Change through metamorphosis happens by itself and interference is unadvisable. Many zoologists and biologists believe that the organization of life retraces the evolutionary development of its species—at least in a general way. This idea was first suggested by Ernst Haeckel. Now, with internal photography, we can see that it is not exactly repeated in all details but is at least true in principle.

If we transfer this idea of preplanned transformation from butterfly to man, the evolutionary development of mankind is a different subject. I deduce that one can see parallels of the butterfly's metamorphosis in many other species. Admittedly, we can never be certain how suggestion affects the cells. We can only reason that inside the caterpillar's body, the cells must reorganize themselves quickly as far as idea and design are concerned when the time comes. This confirms that ideas suggest viable changes involving radical transformation. If one had never seen the metamorphosis from caterpillar to butterfly, one would take them as two distinct species. How many other species have transformed unseen, resulting in no clear link between the stages and leaving us to think they are a different species? Now we know that radical changes do happen resulting in a transformation of design that is immediately functional.

I believe the Universal Principle repeats itself on all levels, in many parallels and analogies, and we are part of the spirit of the universe. Although our own technological creations are only artificial contraptions that require outside energy and human direction, the same ideas of value are also the cause for their existence. The same principle should repeat itself everywhere—if there is only one principle without exception—namely suggestion of value.

One of the most important steps mankind has taken is the development of sources of energy to drive engines. The evolution of the internal combustion motor is one example, and the electric motor another. These can be compared only with organic evolution, because suggestion of value is common, and they are interesting because their evolution takes place in the open.

Inventors in Europe succeeded in building small engines as experimental prototypes, but the motor only became mass-produced after it was considered reliable. Later, similar engines were mounted on flying contraptions, and world travel became affordable to the masses, opening the borders between nations. The world shrank and technology brought people closer together. The world has become a global village as we have come face-to-face with cultural and ethnic differences. If one is open-minded to other ways of living, the world is more interesting. If one is narrow-minded and intolerant, one can fight over values. But the development of nuclear energy made serious war a hopeless situation, forcing us to turn from ideals of heroism as suggestion of value to peaceful cooperation and mutual tolerance. This is a logical development of a change of ideas of value in the evolution of consciousness towards a more tolerant human race. When commerce replaced war as a means of acquiring goods and wealth, we also had to change from worriers to workers and businessmen. This will take time to become fully implemented.

With the invention of telecommunication and the computer we have achieved a form of artificial transcendence on a conscious level, with worldwide connections through the Internet. This is perhaps a parallel, comparable to mental transcendence which unconsciously connects us with other minds in sleep and dreams. But this unconscious channel is not under our control. Why it cannot be is understandable: It would be counterproductive for the separation of values which evolution has laboriously created. Evolution wants to create differences in value and transcendence wants to pass through those separations. This means that in principle technology is breaking down separations that nature has created.

It is difficult to compare transcendence on the unconscious level with the technically achieved information network. But there are parallels, such as the idea that in dreams we seek information and search through knowledge unconsciously. And just as Internet users search actively for whatever they have on their minds, people with physical needs may dream about what they have on their minds.

Let me return to the subject of evolution and the question that has eluded man, namely from *where* are the decisions for transformations coming? We know that the demands come from the combined experiences of countless individuals. Obviously, the opinion of single individuals can not count, but a large majority would have influence, and that only after a long time span confirms that change still is necessary. I do not discount that physical differences among mankind have occurred by genetic variation from adaptation to conditions which resulted in diversity of external appearances, like physical size, skin color and culture. But the basic design of the human species has not changed in thousands of years.

The explanation that we all descend from one single mother is genetically impossible. Mass suggestion is more likely. I think that dramatic events from emotional experiences of a large group of ancestors are the reason for transformations. By default of all other ideas, mass suggestion is the only explanation left, and it fits into a single principle of innate suggestibility to value. In fact, in the history of mankind there are many examples of mass suggestion that have caused social and moral transformation. However, physical changes in design did not occur in historical times, but mental changes have. Why would I believe that from the effect of changes on the social level one can draw conclusions that organic changes could result from mass suggestion on the evolution of species? Because ultimately everything is suggestion: It is the same process whether on the mental or physical level and the same principle that repeats itself everywhere. If there are unproven details, since the pattern is already known, these details must fall into place.

Major transformations must have occurred during geological upheaval, when emotional pressure was frustrating whole regions and telepathic suggestions spread among small populations. During such extrasensory highs, psychokinesis is a possible explanation. Agreed, this is only speculation based on suggestibility, that suggestion of value should also physically change structures of the following generation. The critical point is a change to meaningful organizations that fits the demand.

Because all the relevant branches of science, such as biology and sociology, show the influence of suggestion, all present knowledge appears to converge to a single principle. In medicine the placebo effect also means self-suggestion; in psychology it is called catharsis; in philosophy, ideation. All these discoveries mean the same—self-suggestion has creative influence on the unconscious. I never said every detail is known, only the general principle is. The ultimate conclusion is innate suggestibility and some direc-

a superior dimension of meaning that has influence on the world through subconscious channels. This superior dimension of meaning from another level keeps itself separate from direct involvement and guides evolution by indirect suggestion only. If conservative-minded scientists do not like to jump to conclusions, that is their privilege. Nature, jumps to new dimensions and levels, in which values undergo drastic changes, creating a stratified reality.

In a general consensus, however, accepting the synchronicity of events, the events reveal a pattern of action in which the individual human or any particular species is not the main issue. It is the process itself that could be the reason why evolution is never complete. Individuals are expendable and may be sacrificed by nature for goals that lie outside individual self-interest. One might speculate that many individuals are used to foment conflict, to keep the process of evolution from becoming stagnant. It seems that nature arranges for conflict in order to make the world an interesting experience. In fact, when individuals become stagnant and too secure in their ways, they get a jolt to snap them out of it. Besides, if one should lose interest in life, they will be removed from the stage and replaced by someone who is interested in playing the game of life.

Such is the overall impression of the dynamics of the value-creating process. Nature not only appears to use humans, it also seems to use catastrophic events to stimulate reorganization. Many reasonable people have already arrived at an understanding about nature's intentions: We are puppets. This is, however, a negative perspective. The individual has some input into free choice, and events in the world are affected by our combined actions. The explanation to this ambiguous situation is that suggestibility stands behind free choice. Fatalism and predetermination are incorrect, ancient ideas, caused by pessimism.

Consider the phenomenon of sudden swarming in insects, such as locusts when they simultaneously unite and devastate entire regions? Or when mice in Australia become aggressive? Mice are fearful and normally hide, but they suddenly become fearlessly united. Cats run for their lives, because mice jump on them. To call these events mass-migration avoids the cause—namely mass suggestion. Since animals cannot speak, telepathy connects them.

We know what caused the Mongols to attack and conquer half of Asia and Europe. Verbal suggestion in speeches, suggesting value, made by charismatic leaders like Genghis Khan, caused them to unite to conquer the world. The same suggestion of strength in unity causes other people to rise and assert value, until the suggestion fades or they run into opposition and suffer defeat.

Let me switch to another subject, namely involuntary transcendence. We all have an unconscious connection to this source of information and wisdom, which comes only occasionally to our conscious awareness as a clear dream. It appears we all transcend frequently in nightly dreams for necessary consultation and information. We need this exchange of information for our sanity. It keeps us connected with the source of our being—the unconscious. We are actively involved in dreams through participation in symbolic and visual play-acting, perhaps preparing for coming events. It is also a proven fact that if deprived of dreaming, we soon become confused and then turn psychotic. Since one learns best by doing, in dreams we are mostly doing things, or at least we see this happening. Seeing and doing are believing. They are the most effective use of suggestion, and they are used by the unconscious.

Some speculators, including myself, are beginning to think that everything in the universe is connected. Yet the connection is mostly kept from our conscious awareness, and uses channels that ignore the separation of space and time. Our conscious individual identities may not want to consider it, because the individual wants to feel autonomous and free. But an occasional experience of transcendence makes us aware that there is more to reality than our senses perceive.

From such experiences it appears that nature has already created further levels and dimensions above us. The next higher dimension appears to be on a different mode—on a mode of being where the four dimensions which separate physical values do not apply, where space and time are irrelevant. At least this is what transcendental dreams reflect.

I deviate from the current idea of evolution in many details, but in particular from the forced, violent interpretations of Darwin, Hobbs, Nietzsche, and Heraclitus. Force itself awakens counterforce and brings out hate and anger in people. These are negative emotions that are known to disorganize the mental functions. Positive feelings of value, free of hate and violence, are the climate for creative ideas. But force and violence could at least clear away obstacles and afterwards new ideas can grow. After a forest fire new things will grow, but the fire itself only destroys.

Creation happens by organizing ideas of value in a positive state of mind that is open to suggestions, which makes the idea of creative conflict a mistaken association. I personally dislike competition as pressure because it is unwanted force. Only by organizing ideas in the unconscious mind are creative solutions found.

All higher evolved beings are organized, which means indirectly that some fairness must prevail or the organization will disband. The general direction of evolution is to further organization and to foster greater self-awareness. This demands fairness and reason among the members to keep them agreeable and peacefully disposed. This also applies to human civilization—unfair governments will not long endure.

The idea of perfect justice is elusive. To call our legal courts "courts of justice" is misleading. We should call them courts of reason, because while reasonable compromises are possible, equal justice for everyone is not. Intelligent persons are reasonable and do not require courts. Unreasonable individuals certainly require laws and courts which must exist to prevent anarchy. While most of us are tolerant of some digression from the truth, and hope that people will avoid committing unreasonable acts. Moral theories already exist, but are observed only by reasonable persons. One must first be reasonable in order to understand that to respect the law is necessary for a civilized society, and that it is not stupid to consider others' rights and feelings.

Clear dreams are meant to provide us with information we should consciously think about, provided we can understand the meaning of the symbolic language. One night, while trying to finish this chapter I had just such a dream. I dreamed that I was assembling a wooden model of a building, larger in scale than a normal model. It was a two-story home, about one meter square and high, with many windows and doors. I tried to put an addition—an ornamental gable, that looked like a dormer—on the front of the roof. Although I could see it did not belong to the simple lines of the design, I still tried to glue it to the roof, to see what it might look like. But the glue did not hold and the dormer also seemed to dissolve. So I gave up on this extra adornment, thinking the design did not need it, that it already looked complete.

I awoke with clear recall of this special dream, not knowing what to make of it just then. But later in the day I began to understand the meaning. The model of the house, meant the "model" which I had built by speculating about the meaning of reality, while the entire dream referred specifically to questions of morality, because the roof meant higher ideals. It could be that I was blind to the existence of a moral intent in evolution. Perhaps Plato's absolute ideas of goodness, justice and truth may exist. I never completely excluded this possibility, since moral values are important human values. Even if the lower levels of nature do not have meaning for justice, evolution reveals that nature is evolving toward a higher levels of consciousness and tolerance. But the dream seemed to tell me that the model of reality I have outlined does not need a moral addition. It's already a value-creating process moving towards tolerance. The world is evolving from quantity to quality, from simple to highly organized and self-conscious beings. This model is also about organization. It is a metaphor of nature from a human perspective in which human values are a small niche, in the particular experience of a species; in a larger reality of countless separate experiences, the general purpose is to experience multiple meanings. Any experience of meaning is something, compared to the infinite void.

One could say this model is a philosophy of reason, a system of thought, based on the experience of life. Compare this to the scientific model, which disregards experience as subjective and wishful revery, and has based itself on physical proof of a supposedly objective level of unemotional matter. This concept is unreal, because matter and energy are highly emotional, and the objectivity is caused by a slower mode of reaction of material objects. It is true, however, that materialism takes the physical world as it presents itself to the senses, which to unquestioning minds is enough, while for thinking individuals reality is a phenomenon of suggestion, a fabric of ideas, close to what idealism implies. This pattern also repeats itself in evolution; it is an evolution of ideas that suggest meaning. In actuality, the physical phenomena is temporal. It organizes and disorganizes, thereby transforms itself, but the nonphysical essence remains permanent. This essence is conscious and suggestible, independent from temporal transformation and reorganizations. It transcends physical structures, because it is part of the immortal consciousness we are trying to understand.

I can only see one purpose for the phenomenon of the world: To experience interesting lives. Many think life is a test of moral perseverance against the temptations of the world, but reasonable men agree that the world is set up to cause temptation and conflict. Various religions admit that justice is "not of this world," and believe there is another level where true justice will be done for evils committed in this life. It will happen on a final judgement day, in a finite moral universe. But I think this would be unfair. This world is punishment enough. People are already born unequal, committed to roles in the world theater that they have not chosen but were suggested in their heads, from somewhere. And then we shall be punished for all eternity for playing roles to entertain everyone, including ourselves. This old moral world view is based on a flat Earth concept, unaware of the larger universe in which we are deployed by nature for its purpose to provide interesting experiences that benefit all of us indirectly. We may not always agree that life is fun. The Universal Principle claims that suggestibility to value is the essence of reality. This theory may be close to the real truth, assuming that life is a conscious experience of meaning, particularly since the discovery by physicists that pure energy behaved as if it were conscious of being observed. Suggestibility is elementary to all that has evolved by organization.

6

Of Science and Technology

OUR KNOWLEDGE OF NATURE HAS EXPANDED vastly since the Renaissance, the time of Galileo and Kepler. From studies of the ocean floor we know that our planet is alive, is changing the shape of its skin, and is proving that nature is unpredictable. Occasional volcanic outbursts remind us that we are only deceiving ourselves if we want to believe that mechanical laws can explain the internal processes of nature. Recent discoveries in astronomy confirm that out in the universe similar unpredictable events occur. Galaxies and stars behave as if they have emotions, nothing like our mild human emotions, but more like the impulsive behavior of territorial animals, invading and absorbing each other.

We humans are a step up from nature. Our impulses are tamed and we have learned to control our emotions and in calm moments, think ahead. It would seem appropriate that we should occasionally try to envision the direction that our civilization is going. Let us step out of our preoccupation with creating values and think of the larger pattern of the world.

Nature is an emotional process that frequently disorganizes and reorganizes itself, and saves information in the genes from which it can regenerate. This natural pattern is discontinued in mechanical organization. When we humans began to improve our life-styles, we became impatient with the slow organization in nature and have tried to improve our environment by acting on it. Because sorcerers' magic did not work, we began to experiment and use force. First we used muscle power and ingenuity, then we used heat to melt and reorganize substances by forcing chemical synthesis. More recently, we have used electrical energy in atom smashers to force our ideas on nature.

After years of thinking that the universe is emotional, at the age of seventy-one, I have concluded that the scientific method is negative. Science is based on skepticism, on questioning everything, and on believing nothing until it is proven. To prove something means that the experiment and the results must be able to be repeated exactly under the same conditions. This may be possible with physically stable objects and materials, but this is difficult to accomplish with the genetic diversity and the suggestibility of more highly evolved individuals. In a general way, and in principle only, one can establish a measure of certainty nearer to reality, but the perfect repetition of living beings is not possible in nature. Admittedly, scientists are individuals and may have a positive disposition and open mind toward life that counterbalances their negative scientific skepticism, but in principle science is negative and physically based. In addition, it has become necessary to specialize, resulting in further estrangement. This skepticism is applied to each other's conclusions. They also distrust each other. All this means is that scientists, educated to disbelieve, are difficult to convince. To disbelieve in the meaning of the universe, in causality and reason—that leaves measurement and calculation as the only dependable source. With this, they are limited to physically stable materials. Therefore the name of scientific materialism is correct.

If one is trained and used to thinking in mechanical terms, it is like a dogma. In fact, the belief that the emotional excitement of suggestible quanta heats itself up when it joins into whirlpools and creates light (the same light that physicists treat as mechanical), they cannot believe it because gravity is their fixed idea. I see this light instead as the light leading to conscious reason.

The generalities of emotional energy

Technically, not much would change if a new model based on emotion were to be introduced, except that electricity and light need not travel. It only transfers emotional excitement, either through empty space or through metal wires. Now I understand why emotional persons like Uri Geller and Thomaz Morais could cause metal spoons to bend. They excited the materials emotionally. Indeed, the entire psychokinetic effect is understandable if you think of it as an emotional frequency that happens to match the material level by synchronization.

Fortunately we are not normally that excitable, and therefore our world remains solid. Things keep their value and identity as we know them. Nevertheless, if nature *is* emotional, the mechanical concept is the wrong perception of reality and we were misled into a dead end by the mechanists. There is still time, though, to swing around and go with nature, but we must rethink all we know about the nature of light and electricity.

Some of the natural events on our planet can be explained by thermodynamics and kinetic effects based on local gravity. Such natural events are limited, however, to the material level and to stable substances. When electrical forces and subatomic transformations are involved other explanations must be found. Gravity fails to register as physical energy on instruments, implying it is probably a suggestion of value. Electricity, although immaterial, has some measurable physical attributes, yet it is also affected by emotions and therefore is not actually mechanical. And while electricity can be created artificially by magnetic fields in generators by causing tension, electricity in nature is caused by tensions in the conflict of values of which we are examples.

But while science may use mechanical generators to create energy, nature uses emotions. Nature is organized by ideas of meaning, ideas that suggest value to create exciting worlds. The critical point that should raise doubts is the presence of innate order and organization in nature, indicating that it is organized from within by intelligence. But science is accustomed to experimentation and can not see that "meaningful evolution" is guiding the organization. Experimental science is only discovering external alternatives, not the real cause, implying that experiments in laboratories have little in common with natural organization. It means physicists are mistaken if they think they can understand how nature is organized through mechanical experimentation. Innate suggestibility from the very beginning is the actual cause. Let me ask the question: Are we humans meaningful participants in nature or meaningless mechanical contraptions? I am certainly not meaningless nor do I think others are. This goes for all of nature.

I do not believe that atoms can be forced to fuse, but that they have organized by suggestion of value on their own. Experiments that suggest that atoms can be charged to jump to a higher level are laboratory-based artificial creations that have no relevance to nature. One could think of electrical wave motion as emotion, and relate it to the larger picture of an emotional universe. It seems possible that cold fusion is a better choice, although the gain of energy seems minimal. Besides, we may eventually understand more about energy.

But, for practical use and on the time scale within which we humans live, the cooled surface of our planet is stable enough to be treated as lifeless, and mechanical methods can be applied within the range that substances can be trusted. For such limited purposes mathematics, using mechanics, are useful as a technology to invent machinery that could be compatible with nature to make our lives easier. But these would be only mechanical, lifeless contraptions without consciousness. Just because a few artificial substances have been forced together into a chemical synthesis that never existed in nature, does not mean the principle of the universe can be explained with a formula. Quite the opposite! Nature is driven by emotional energy and organized by innate suggestibility to value which is mysterious.

In fact, the experimental method that began with Galileo has developed in the last three hundred years into a mechanical technology that has little in common with conscious beings. Conscious nature is organized from within through a long evolution and tests made with instruments tell nothing about the existence of consciousness. According to experimental science, consciousness should not even exist. We should not be alive.

I have mentioned that the idea of gravity is unreal and it is useful to speculate occasionally where we are going. If we humans and everything else in the universe are derived from a common source or essence, then we should all have something in common—emotionality and suggestibility, which together form consciousness. Nature is the result of organization of emotional energy.

Organization has always interested me because it explains many existing phenomena. One of them is loyalty, which is necessary even in marriage. Loyalty is expected from members of any organization. In fact, without it, organizations can not be sustained. The human body is a cell-state which functions well if value is suggested to its participating members through their cells and their atoms. But if the suggestion to inspire value should be weakened by loss of interest or should turn negative, the organization will malfunction because trust and value are lost. This explanation can be applied to any political or natural organization including the stars in the void, and it explains why the planets are loyal to their mother sun. It also explains why children follow their mother who loyally cared about their survival. Yes, there are differences between stars and humans, but in principle it is emotional attachment and loyalty that holds things together, not gravity.

Although scientists are basically skeptical, an emotional explanation of nature is nearer to reality than a mechanical concept. I may not have a scientific education, but I do have practical experience with automated machinery and I can assure you that computer-guided robots and electronic vision are totally meaningless for themselves without humans to guide and use them, because without consciousness nothing will ever work—not human beings and certainly not mechanical contraptions.

The logical deduction is that all matter is a manifestation of waves—in fact a synchronization of waves—from which it is conceivable that the vibrations could have organized to create meaning. At this point a new model should have been introduced, particularly when it became clear that organic life is incompatible with a mechanical universe. Instead, living reality was declared a freak accident of nature and extrasensory phenomena were ridiculed.

The alternate idea of emotional energy is supported by the excitability of human beings, who get themselves into emotional states, then feel heat waves raising in their bodies and become emotionally charged. In highly frustrated individuals, their mental states of emotional excitement can effect materials. It is incorrect to call this phenomena psychokinesis. Emotions are involved, not kinetic energy. If we all are from a common substance or essence, the emotional nature is carried through the entire process of creation as organization by meaning. From this perspective our psyches could be part of the greater psyche of the universe, and it is naive to assume the world has accidentally happened.

If human reason is only speculation and science needs working proof, I shall try to provide it. But, I know it is difficult to convince ingrained skeptics who have closed minds. Fixed thinking, demanding mechanical proof in detail, was apparent when the concept of local gravity was extended to universal law. Galileo, Newton and Einstein believed that gravity is a universal force among the stars and explains everything in the micro- and macro-cosmos by mathematical formulas. In fact, universal gravity does not exist—the planets follow individual orbits by their own will far beyond the range of local gravity, defying mechanical law. This has not stopped physicists from pursuing this mechanical concept because it has mathematical symmetry and elegance.

Following gravity, they were led to the conclusion of a necessary Big Bang Theory of the creation of the universe. Frantic calculations are still perused to try to integrate universal gravity with quantum wave theory in order to find a mechanical formula for the grand unified theory of the universe. This obsession for a mechanical formula has been misleading physicists ever since Galileo suggested it until we found out there is no gravity in space because astronauts float freely around. This seems to confirm my idea that innate suggestibility can explain the movement of planets in the cosmic theater. If the planets were individual beings with identities of their own (as astrology believes) and orbit loyally around their mothers, then they could move nearer to them or farther away at will without falling into it or becoming lost in space. Besides, the outer planets—such as Uranus, Neptune and Pluto—are so far from the sun that gravity is unlikely to hold them in orbit.

Another example is the occasional comets straying through space and passing near to the sun without being drawn into it. Because the sun does not attract or push away, it only affects the comets emotionally because some of the ice is melting. Emotions are obviously what the sun sends out. The sunlight excites atoms and molecules when it falls on objects and beings, causing them to become excited and heat up to a relative degree. This also explains why space is cold, although sunlight passes through it. In principle, light affects things by innate suggestibility which awakens materials emotionally from within through suggestibility which is also the case with psychokineses when emotional humans excite materials.

Materials are not considered autistic when compared to humans because they do not object to being used. They are self-absorbed in their own value and on a slower mode. Autism has been explained in more depth in Chapter 4. Unresponsive materials are the type of reality the experimental sciences consider worth studying. It must be stable enough to be measured with instruments one way or another. This excludes conscious and capricious beings—the material level suits science best. But, a substantial part of nature above the physical level is organic, which cannot be totally ignored. Science wants to study this unstable part as well. Therefore, the sciences in themselves are divided into the hard kind, which use measurement and mathematics in its experiments, and the soft sciences, which use reason in addition to measurements. Science is divided into specialized fields in which even the experts often disagree about details and become absorbed in their own expectations. In fact, scientists are just human, and their belief of objectivity is just a suggested belief.

In order to inform myself about the latest theories in nuclear physics, I read a recent book by a renowned physicist who has many academic qualifications. This author wasted many pages attacking philosophy as the curse that has mislead science. He said that speculation leads to ideas that are assumed true, and outrageous claims are made without experimental proof.

I took this qualified opinion in reverse order—that detailed proofs are misleading unless they can be integrated into a larger system that has different levels where there is a logical explanation. How many proven theories have already been refuted? Will gravity be relegated to a local role with a special purpose in the design of nature? The claims that are made by my philosophy, such as a suggestible psyche, are not outrageous in the larger psychology of nature. I have great admiration for experimental science, but its attacks on speculative reason and philosophy demand rectification.

I took the next step, implied by psychokinesis, to assemble a theory in which suggestibility to value is the cause for the self-organization of reality. This cause is also backed by electronic parallels when waves create phenomena that seem to suggest reality as in holograms. But my theory came from transcendental idealism, an ancient philosophy. This was the nemesis of scientific materialism, sometimes even called spiritualism, as if this would refute transcendence. Spiritualism in philosophy is meant only as the spirit of nature that repeats itself everywhere in principle and is believed to be the cause behind the organization of everything, starting on the nuclear level. In the domain of nuclear physics, this is the stage where the organization of waves as materialization begins and expresses meaningful ideas.

The manifest world of material things starts in the atoms, where undifferentiated quantity organizes to become differentiated quality into ideas of meaning that have specific value and assume identity. Quantification would only mean more of the same thing, but it is the difference in quality that makes the world interesting. That should not imply that on the subatomic level only quantities are found. Mental suggestion may already exist, although it is difficult to determine because suggestion is invisible. On this level, emotions are raised to recordable vibrations of emotional waves into wave bundles. They are also called energy particles, with many names given such as mesons and quarks, that imply unpredictable behavior. I do not question the existence of particles, only that they are not mechanical, but emotional entities.

Perceived reality is a phenomenon of wave motion at high speed. Every child learns this in school, but is not taught that the motion of waves are caused by emotions and become organized by meaning. Waves are supposed to arrange themselves accidentally. If they are intentionally organized, then the question arises, by whom? If there is someone, where does this being come from? The existence of a single being is unlikely, because the process of organization is a search for meaning and identity in a massive scale.

Visionary seers

Transcendental seers, in particular the Zen masters, say that the ground of being is total emptiness, yet conscious. Because nature is undefined consciousness and still evolving, this eliminates the idea of a creator, leaving only innate suggestibility as orientation to organize nature. This is not just an idea—innate suggestibility really exists and can be demonstrated during hypnotic trance. In essence, the phenomenon of reality is caused by innate suggestibility, which is emotionally excitable in primal energy with inconceivably greater force than on the organic level.

Because the organization of nature is meaningful, intelligence must be involved. That is why an intelligent spirit or mind is assumed to exist, and, some think reality has evolved from conscious mind fields instead of quantum fields. Agreed, this is speculation, but is there any other way to make sense of this except to call reality chaos? That is all that the physically grounded sciences can tell us. In their opinion, the world makes no sense, and existence is meaningless suffering. This is similar to the lessons of Buddhism.

It is useful for science to remain dedicated to discovering details that will help technology to improve human life. But it is a naive generalization to argue from the physical dimension with mechanical rules to explain the emotional universe. There is a need for objective knowledge to establish standards for the classification of physical data. Science is well suited for this purpose, but for a larger comprehension of the diversity of nature, it is necessary also to organize psychological details into a meaningful whole. For this, speculative thinking is required and imagination is an asset.

Considering all perspectives, it is now possible to form a new theory of reality. Science has provided many details at the physical level. To have it make sense, speculation has integrated it all in the conscious, pluralistic psyche of organized individuals. If one is open-minded to many phenomena, it is possible to believe this idea.

But let us go back to what is still the position of most conservative scientists—that true knowledge is based on proof. Science wants to establish true knowledge, which can be verified anywhere with the same conditions. That would be fine, if reality were stable. But since it is not, subjective knowledge is divided into a class by itself, comprised of unstable beings which can be ignored as unimportant because they do not fit into any mechanical calculations and a material universe.

In the understanding of *real*, a mistake is made. The subjective, nonconforming beings are not counted, although they form a substantial part of our world.

If one wants to understand reality, even as a late-developed level of consciousness of the material universe, it is wrong. Consciousness and orientation are already present in subatomic energy and necessary to explain the organization of the physical part into the organic mode of mobile beings. The separation of objective and subjective knowledge is unnatural; it is an artificially created dichotomy. If there is a separation, between naturally organized and artificially assembled things, it began with technology. Natural things have slowly evolved by themselves, but artificial man-made things are reorganized to suit human needs. In the process of manufacturing, the autistic character of materials remains undisturbed. Indeed, in respect to the stability of materials, technology has created an artificially organized world where the parts respond mechanically, typical of repetitive functions. Absent in manufacturing is a new identity of value consciousness that is naturally acquired through gradual evolution. Artificial things are mechanical assemblies without a common identity.

In practical terms, the collection of data from nature by experiment is the genuine value of science, but mechanical theories based on mathematics are added abstractions from reality. Experimental science began through trial and error after the Dark Ages. Chemistry was known as alchemy in the later Middle Ages, when in sinister laboratories, chemicals were identified and mixed. In time, experiments were based on measurements and quantification. But to make quantification comply with the principle of the universe by calling God a mathematician, was left to Einstein, Rieman and others in the twentieth century, when mathematics left reality and became abstract.

Let us start with a simple experiment to understand quantification, beginning by establishing the strength of materials for construction use. The experiment begins with several equally machined bars. Each bar is subjected to different forces, from bending under load to breaking, twisting, pulling, burning and decay. Several bars are needed—obviously each is slightly different. From these tests the data is established to be used by engineers in the design of structures and mechanical devices.

The experiment determines quality based on standardization by deliberate sorting and processing of artificial materials that do not exist in a natural state. This is just a basic idea of standardization for the use of engineering. The processing of metals begins from the ground up, through smelting and mixing, casting and machining—so far from nature that mechanization is literally dependant on artificially developed materials. Only after processing to standards are mathematics and geometry useful, and then only on materials which are still subject to decay in time, suffering fatigue by loss of meaning, and possibly affected by certain wave frequencies. We know that technology is a human achievement. No other species on Earth has developed such a high level of mechanization that resulted in space flight.

Technology is for human use only, being useless for the rest of nature. In fact, it has nothing in common with the principles of nature, which have evolved by separation into distinct identities and evolved intelligent individuals who use their intelligence to investigate nature. Some scientists work on expensive projects where competition with others forces secrecy upon them, such as work on nuclear energy, semiconductors, cold fusion and more. For example, cold fusion is contradicting the laws of thermodynamics. Instead of loss by entropy, it shows increase of energy. This means cold fusion would fit a theory for creating energy through positive suggestibility in substances. This may just be a start of a slow beginning in a technology that creates energy based on emotionality in substances.

Although scientists are unconsciously following the pattern of evolution, dividing and reorganizing into greater diversity, they are still unable to create conscious beings. The mostly chemical and mechanical combinations that technology has created are just assemblies without a common consciousness that require outside sources of power and outside guidance for orientation. Many scientists agree that from a mechanical perspective, the universe is meaningless and chaotic. But a new model of reality could change all that and restore meaning.

Philosophy, at least the panpsychistic, idealistic direction, is trying to connect the parts and fragments of knowledge to discover the gestalt and meaning of the process as a whole. For this purpose, the fragments of data can be useful information, but have to be separated from scientific theories and reduced to their essence to fit in with a conscious universe where even the stars have emotions.

I do not even pretend to know more than basic generalities, and strictly intend to integrate a wide spectrum of already-agreed-upon ideas into a single principle in which the details must fall into place. Yet, I have the same common interest as science, trying to understand the same reality, but from a rational perspective, and must pay attention to confirmed data of the experimental science but with full regard to the whole of living reality.

As mentioned, when thermodynamics was the dominant theory, the energy in the universe was thought to be a form of heat that was gradually diminishing, making the world terminal. After the subatomic level was discovered, the energy became electromagnetic, and events were explained by the electrodynamic quantum theory. But now, after innate suggestibility has been discovered as fluctuating emotional energy, the events in the universe should be explained as psychodynamic. Dynamic now means emotionally created energy that not only can decrease, but also can increase, and is controlled by value.

This implies that suggestibility to value organizes the world through meaningful evolution and that psychodynamics is two-sided. Positive ideas create value and negative ideas diminish worth. The general trend, however, is mainly positive because the expansion and reorganization from simple to complex dominates over the negative side. The model of the universe that I have perceived is mainly positive. It is one in which more emotional energy is generated than lost, yet the negative side is always there as reminder that if the emotional mood becomes negative, it will lead to disorganization, but not to finality. This means it is an emotional, cyclical process, responding to experiences in the world that are guided by meaning, and fit in a system which is a sort of value-creating process with many obstacles which can be seen as challenges to provide interesting life experiences.

The belief on which astrology is based, that the stars have emotions that affect our human psyche, is ancient and may have a valid point. Consider that in Babylon knowledge of the universe was based on a flat Earth with the sky turning over it, but people felt an emotional connection to the little lights in the night sky they called stars. The knowledge of stars has expanded in the last three thousand years as many become upgraded to galaxies, but astronomy does not exclude the possibility that emotional connections may exist, even as a part of the psychological side of reality.

In recent times many of the ancient ideas outside physical law and individually different have proven themselves in principle, that a pluralistic psyche must be assumed as the basis of being. Because of the extent of such a vast reality as the universe, it is unlikely that one single, omnipotent entity could oversee or control the process. It is reasonable to conclude that there is a general principle instead of a single God. Philosophy is open to the existence of the larger reality, while science cannot accept metaphysical explanations because they are outside physical limits.

Agreed, there are basic differences of method. Science uses instruments that extend perception. The latest is the computer chip on the telescope to scan the outer universe. But for the past four thousand years there has been a long line of philosophers, among them many with transcendental insight and ideas that have similarity in principle, containing information that is available to anyone and may be nearer to nature than the mechanical explanation.

According to the latest scientific terms, thought processes are accompanied by electrical activity among brain cells. But at this point, we begin to see that thought processes are very different from presumed electromechanical functions—nothing is learned about the content of thought from recorded waves. The brain cells are organic in function, and the response of cells to suggestion is confirmed which indirectly create electrical activity in the brain from the thoughts that suggestion instigates. Additionally, all cells consist of molecules that consist of many atoms and atoms consist of waves created by emotion. This confirms that emotions indirectly create electrical activity in the brain, implying that energy (including organization) can be influenced by thought.

Experienced reality is defined as limits and the important processes are metaphysical and unconscious. We presently think that ideas that Socrates and Plato called immutable have evolved and are constantly transformed as the values change in the continuing reorganization of experiences. Since we are unaware of the inner organization, and only occasionally become aware of transcendental connections, we are left to speculate. Uncertainty is still with us. That transcendence occasionally becomes evident in normal life supports my hypothesis of innate suggestibility. We know that not only other humans, but also animals and even materials can connect to our thoughts. Yet one is never certain to whom one is connected.

Consciousness is certainly shared among all beings and is the only known medium that can possibly connect everything since it is present on all levels of reality. In thought, we can transcend the separations of value and enter the realm of the uncertain and imaginary where nothing is sure. We are also the only species that applies speculative imagination in fabricating artificial things. Although many species also improve their environment, it is far less than we have done and will continue to do, hopefully in harmony with nature.

The chance of improving our physical life is made possible by the use of tools. Examples are the recent computer-guided automated machines that theoretically make no mistakes unless they break down. We should not be surprised that computers, without conscious memory, understand nothing and know only wave patterns. The conclusion reached by some persons that computers resemble thinking humans, is a doubtful expectation. Computers organize knowledge on a digital basis, different than do we, meaning they should have no idea what goes on, and yet sometimes it seems computers are suggestible and might know something.

Practical technology

We humans could perhaps be content with life as it appears to the senses if we could be totally unaware of other dimensions. Then there would be no issue. But since this condition never exists, we are aware of many exceptions to the rule and continue to think about other possibilities. Rational thought is like an additional sense that can transcend the barriers separating levels.

Transcendence in logical thought became the great achievement of the last century. I am not referring to extrasensory experiences, but to the transcendence by rational thinking that led to many discoveries in science. This happened because of the inquisitiveness or even better, the suggestibility of the human mind. Consider the case of scientists who have used instruments such as telescopes and microscopes, making it possible for us to confirm that other strange levels also exist. For instance, we would not have known about the vast extent of the macrocosm or the depth of the microcosm, and we would have been left with unconfirmed ideas. The natural sciences, with the support of the latest technology have lead to the breakthrough to new dimensions which are not exactly physical any more. Knowledge of the existence of other levels should not affect our lives directly nor disturb our minds. The average person does not care one way or other, except for the consequences of the use of knowledge by technology and exploitation by commerce, which profoundly improves our life-styles but have devastating side effects on the health of nature and man. It is lucky that countless protective separations are wisely build into the system of nature, limiting the damage of any kind of chain reaction to sections.

The existence of separation is also recognized by physics, which understands nature as a structure of levels called stratified stability, generally impermeable and limited to suggestion. The stability is conditional to stable materials being content in their own simpler self-images of value. However, nature is much larger than the material level, and not quite perfect in upholding separation and stability, not even on the material level. There are always exceptions and cracks in the veil of secrecy, caused by discontent of beings with their allocation of roles in the scheme of things. These are beings who either are searching to find ways to climb higher, or become disillusioned and degenerate to a lower, less self-aware level where they no longer care.

Before the existence of levels was known, scientists and other logical persons believed the entire world consisted of one level—the perceived and tangible world. All else was called the fabrication of overactive imaginations and superstitions from the Dark Ages. But because some photographic plates which were covered and stored in a dark drawer together with a few rocks, became foggy, someone wanted to find out how light got to them. For most people this would have been an insignificant event, but not for Henri Becquerel. He kept thinking and talking about this phenomena and concluded that inside certain rocks there is a source of light, radiating invisible energy and brought this almost unbelievable idea to the attention of a few who listen. Pierre and Marie Curie explored his theory and undertook the task to find the source of radiation by refining enormous quantities of pitchblend until the source of energy was condensed in a small quantity of radium. It was the radium that radiated invisible energy and acted like dark light. This led to the general deduction that locked inside all atoms is pure energy, which in low concentration is quite harmless, but in a pure, undiluted state emits dangerous radiation. The Curies concluded that an atomic substance such as uranium isotope U_{235} naturally decays and is extremely destructive to living cells while most other atoms are healthy and content.

Before the discovery of atomic decay, many persons had independently begun to experiment with electricity and invisible energy. In his work with invisible wave energy, Wilhelm Roentgen discovered that X-rays can pass through materials and also through human cell tissues, confirming that materials could not be solid and thereby implying that physical reality could be a wave phenomenon. Ernest Rutherford discovered through experiments that atoms can break up and concluded that matter can transform into energy. But the other equally logical possibility, that thoughts can create waves that transform into material reality, did not seem to have many followers. This is metaphysical speculation and leads to idealism, implying the waves could organize by ideas.

Even more to the point, during this period of discovery other invisible effects demanded an explanation. J.J. Thompson proved that atoms, apparently made of matter, must actually consist of electromagnetic wave energy that is normally in states of balance and by imbalance, cause electrical current. The atom was thought to have a center with a mantle of electrons around. These electrons have some freedom of movement, which accounts for the flow of electricity that is always trying to stabilize itself to achieve balance. In general, the idea that atoms are bundles of electrical wave energy caused by electrons spinning and pulsing at high speed became accepted as causing the impression of materiality and the perception of solid substance. From such deductions it was inferred that atoms are separate units of empty space in a high speed of vibration. The synchronization of waves causes the solid appearance of things, thereby implying that all physical things are a phenomenon of electromagnetic waves, creating the illusion of the physical world.

But, the discovery of electromagnetic energy did not lead to the final conclusion that everything is a phenomenon of empty waves, as it can be expressed it in simple terms in the present time without causing disbelief. On the contrary, scientists then believed that materials were the real state of being, and energy was a form of materiality. Certainly, as careful scientists, they did not jump to unproven ideas. Besides, materialism was the current philosophy, financially supported by governments on which scientists depended for their livings.

The discoverers thought of themselves as scientists, believing that they had explained everything by mathematical equations, finding formulas to bring them into the established system of physical laws. They obviously believed that life and the world were physically real and electricity was a form of matter. As such, the world remains materially and physically real and not an idea in the mind of God, as George Berkeley thought. In fact, there is not a great difference between the evolution of consciousness and the slow organization of the world by ideas. This resulted in the tangible reality that feels real, which I now assume the phenomenal world is, in which all participate for the purpose of experiencing meaning which then was not a possible notion.

Presently, our technology is based on electricity. We are quite used to electronic devices and wave motion as a source of light. The worldwide communication systems, all of which have been introduced in the short space of one hundred years, depend on wave organization and distinction. This should have affected philosophy and the general perception of reality, but materialism has remained dominant and all else was ignored.

In spite of the accepted fact that atoms are totally empty inside and electromagnetic energy is invisible, most physicists and chemists continue to think of the world in material terms, based on the physical model from the past, and then ignore the implication which the organization of wave patterns into meaningful things suggests. In fact, the thought of organization contradicts mathematical order and physical laws derived from calculations.

I don't know why the general public was not ready yet for an emotional concept, not to mention the philosophical consequences that would allow for transformation and transcendence, which is necessary to explain many phenomena that still are called impossible or delusional. But materialism was established and a change in thinking has been resisted. Perhaps this conservative position is a naturally occurring delaying action caused by the need to defend existing identification with values and to slow evolution. Stability is necessary to establish order for a while, but the idea of progress is to speed up change. Technology is a direct cause for runaway mechanization and mass production. It is not the solution for human survival. It may improve the standard of living while indirectly destroying the ecology that has been organized by slow evolution. We humans could become standardized and homogenized if mechanization had its way, but people, in particular labor unions, have revolted against becoming obsolete by mechanization. In any case, nothing has been done. The material model with mechanical laws is still in place, but its influence is waning.

At the other extreme, for people such as fiction writers, the transformation of matter into energy and the reverse, energy into matter spawned science fiction. These writers describe events far out from actual discoveries. This has caused the return to fantasy, because readers began to lose interest in mechanical contraptions, that are limited and final.

It is understandable that many intelligent persons feel it necessary to caution against overreacting to the importance of the psychic and occult, but there is a middle ground. Skeptics may categorically reject all implications of intentionality or the existence of purpose and meaning in nature, but their position is incompatible with facts. At least an open mind allows many sides to be considered, including the possibility of consciousness as a self-orienting, general principle.

Although scientists are trained by scientific method to be skeptics and are supposed to avoid speculation, they may still be idealistic thinkers who have devoted their lives to research serving mankind. They may be unconcerned about wealth and luxury, and morally may be more selfless than many philosophical idealists. All this means there are no laws to define self-conscious individuals. We are thinking beings, able to recognize changing situations, unlike mechanically programmed robots.

At least experimental science and technology are in touch with reality. Theoretical physics, however, seems obsessed with inventing systems to calculate the impossible without relevance to the real world. Mathematics can be true in itself, yet unreal.

It is the real world that is capricious, emotional and changeable. Others, including Bertrand Russell, author of *Principia Mathematica*, eventually gave up and called mathematics "tautology" after spending most of his life to trying to fit it to reality. It is counterproductive to talk negatively about a subject like mathematics. Only positive explanations can make people see things in a different light. From the ranks of scientists grows opposition against abstract systems, against the Big Bang cosmology and against gravity as universal law.

Nobody knows how old the universe is or whether time and space are relevant beyond the separation of value. A group of physicists assume that the ground from which everything has evolved is electron plasma from which waves emerge, forming the electrons that are basic to all things and sufficient to explain why electronic technology is possible. This explains the universe as a straight electronic wave phenomenon. Of course, it still is a mechanical theory without conscious orientation. It is much simpler than a Grand Unified Theory with four forces, with super gravity and a wrinkled space-time shroud with worm holes through it to account for transcendance, contrary to a simple principle that was expected as first cause.

Einstein's theory of relativity which uses space and time to connect it with gravity, arrived at a curved and wrinkled space-time continuum of relativity, but it solved nothing. The rational idea of space is just emptiness. It serves as separation between entities, such as waves, electrons and organizations including us. The separation in space requires three dimensions and time as the fourth dimension, which is another separation required by constant change. Space and time are separations of value necessary for a value-creating process in which change is the only constant.

It may seem simplistic, but the pattern of evolution is from a simple emotional cause to very complex organizations, and is open ended. Other persons with credentials in mathematics and physics have questioned the latest developments in unifying all theories with abstract calculations. I am not assuming that they would endorse innate suggestibility although it is already confirmed as present in the primal essence of reality. This reminds me of the problem I encountered with suggestibility in hypnosis: it was without content. This problem was solved by the criterion of value and meaning. Perhaps theoretical physics should also ask whether mathematics is empty, without conscious beings that have purpose and value and thus imply that the universe has meaning. From the behavior of living beings, one could assume existence has meaning to them.

Regardless of how much more sensible the plasma-electron theory is than other theories, it may actually exist only in the mind. In any case, this entire reference to physics is unimportant. According to my own theory, the universe has a suggestible psyche, which is the basis of being. This psyche is emotionally excitable by suggestion and the energy it creates is emotional energy which closely resembles electrical energy, and it is conscious. Emotions create waves and ideas organize waves into meaningful gestalt. That is how it works, but *why* it exists is the question.

The practical application of technology actually agrees with the electrical part of the universe, but suggestibility has to be added. In fact, I see a parallel in human emotions that resembles the pattern of generation of electrical energy in mechanical generators. The difference is that mechanical generators are driven by an external source of energy, but living beings internally generate their own energy. This happens when humans are challenged by the world and each other, thus, raising their consciousness and energy level through interaction. In interactions with the world, individual life energy is elevated to a crest, after which it levels off and in old age declines. From the generation of emotional energy, civilization evolved along the pattern of the dialectical principle proposed by Hegel. A single person may not generate much energy, but a collection of many humans can produce large amounts of emotional energy through mass suggestion. On the human level this is certainly a very different process from mechanically generated electricity, but it is the generation of emotional energy in principle, and is caused by suggestibility to ideas of value and meaning.

Such a comparison between matter and mind may seem far-fetched when only the principle is similar and the physical conditions are totally different, because beings exist in different modes and are separate in ways that make them incompatible. Differentiation is basic for values to exist. That is exactly it. In real terms the modes are spaced far apart, estranged by differences in value, to keep the universe from internal confusion and collapse. Overtly we seem to have little in common, but indirectly there is the emotional connection, which psychokinesis confirms. It is the real presence of innate suggestibility that explains the evolution of consciousness from a common ground by differentiation into the individual experiences of being by complex organizations: This is the principle which applies to everything in the universe.

From the assumption of a common origin we know that the physical and psychic side of experience belong together; that our conscious orientation is subconsciously guided by innate suggestibility of values, and that ideas organize the meaningful things and beings in the world. In fact, even materialism is based on value and organized by ideas. The founders of business and industry have organized people and used ideas to produce things that suggest value to their potential customers. Yet behind this physical development stands the self worth of the individual, his or her identity. When it counts, this will still assert itself and suggest action, not necessarily of true value, whatever value is for the individuals.

As important as the material level may be, it is only a background like the setting on the stage of the theater of life, which must appear real to give meaning to the show that the experience of life involves. The show of life continues on, the players and stage consisting of different entities, change and differentiate as far apart as stars in galaxies, as atoms in the inner space of humans. The universe replaces its actors individually without shutting the process down, yet many actors think that without them, the world ends. From many indications and clues, the replacement serves the same purpose for the individual, namely to provide a new experience when life has lost its value. The individual's consciousness survives and only transforms its gestalt because consciousness is immortal.

Regarding the question of the continuation of consciousness beyond death, I had discussions with medical doctors who said they never saw a spirit leaving a dying person and therefore believe when the body is dead, the spirit is dead too. To this I answered, who ever said the soul is physical? Spirits are on another frequency that average humans, with some exceptions, cannot see. Some clairvoyants sometimes can see spirits. All of us have occasionally seen spirits in dreams, and not only met the spirits of deceased but also of still living persons. Apparently in the dream world, the spirits can meet to exchange information if they desire, implying that dreams are selective. This is possible because in dreams we lower the physical barriers and the mind wanders in wider fields.

Many such ideas have been believed by mankind. Reports of clairvoyant experiences are common in all cultures, and many theories exist that differ only in detail. In the West we have had various seers and magicians, such as Nostradamus, Swedenborg or Rasputin, who foretold the future. Their predictions were confirmed, but could not be prevented from happening. In fact, in ages past influential people believed in the possibility of foretelling the future. Among the seers were the alchemists and free thinkers who believed in the magical and mysterious, especially during times of religious fanaticism, which endangered the pursuit of knowledge. Many people in the present still believe in prognostication, using forms of divination such as the I Ching, tarot, cabala and astrology.

In this century, such claims of extrasensory awareness led to the establishment of a special science called parapsychology. The word "para" means beside or outside of psychology, which means it was more intended to establish that such phenomena are illusion. Yet it turns out that there is something to it—new data has been added to expand human knowledge. The narrow interpretation of nature by mathematical physics as a mechanical process is not shared by all scientists, otherwise parapsychology would not exist.

New model

The new model I am proposing is that from the beginning, if the universe had a beginning, innate suggestibility was present and continues through all manifestations. It can be assumed to exist when fields of consciousness search to be something of meaning and unite to create and to evolve where nothing but fields of consciousness exist in the void. Therefore, the value-creating process makes sense as a colossal organizing process. It is driven by the wish to be something. Anything is more than nothing. Based on this wish of consciousness, we and our special planet are among the greatest successes of self-creation—at least from what we can observe with our perception. This leaves the possibility open that many other worlds could exist.

We correctly call the universe the union of diversity. Ultimately, each separate life experience has meaning either individually or collectively. This opinion is shared by generations of philosophers and some outstanding scientists, for example, two of the discoverers of quantum physics, Werner Heisenberg and Erwin Schroedinger. Each independently believed in a larger reality beyond the physical known level in dimensions we cannot see called spirit levels. I assume they believed that the physical world is background for the experience of meaning in life.

By now it must be clear that the experience of meaning is more than the material background, which is the structure for the initial purpose. This information is confirmed by history and psychology. With psychology, we have reached a point where all data and logic leads to an emotional, suggestible essence as basis of being, which intends to become something, as single cause of the universe.

I believe this, but I also know that scientific materialism is deeply entrenched in our civilization, and am still convinced that consciousness to them is only a by-product of matter and nothing more. When the body dies, consciousness ends. Scientific materialism persists, because most people believes only what they see directly, but common sense and electronic technology must tell them that many unseen waves and phenomena of things exist in reality. The larger part of reality is invisible and only the mind can indirectly grasp the meaning of the whole.

The development of electronic technology in our time has polarized the issue between mechanical and emotional energy. It used positive and negative signals as combination for words and calculations. Computer language has evolved from this, making the development of software a very lengthy and expensive process. Compare this to conscious thought, where ideas organize instantly by meaning because innate suggestibility is conscious and recognizes change which digits do not.

Obviously, the same electrical energy can either be perceived as emotional, or dreaded as autistic mechanical energy—at least on the elementary level. We know, however, that under normal conditions, even mechanical energy can be influenced by emotions, although we do not know exactly what this involves. Nature wisely installs protective separations to prevent the confusion of waves.

Not from understanding but from experiments, without putting ideas of meaning into it, science collects data and technology uses it. This is done mostly for the exploitation of nature, including mankind. Because something works, people rush to make a profit from it before somebody else beats them to it. To maximize profit is the goal of business; nature and people are secondary. Because everyone does it, nobody is responsible. In fact, many actually think this is a law of nature, where all compete for a place in the sun. This is the thought in the minds of the physically and materialistically oriented majority of human beings today.

But this is a misconception. It is not the value of material possessions that we can gather in life, but the experience of meaning that we possess. When a lifetime runs out, all one can call one's own is the self-worth and experience. Our identity consists of the memories we have lived in the passage through life. The wise men of all times have said this before, although perhaps not always using my definition. Many believe we are here to serve God. But if there is such a mind (God), it is standing in a distance, letting the world run itself as it may. This implies that we are responsible for what we make of our lives, using the inherited potential we are born with, and the knowledge we have acquired in previous lives.

It is logical that behind the physical is the metaphysical, the unconscious that connects the apparently separate parts to a meaningful system. All things belong into a collective consciousness that ultimately is suggestible to value. Consciousness lives in everything. It is the first cause and also the cause of all further causes that follow infinitely. Such concepts are as old as metaphysics. The belief in causality is basic. Only the newly invented mathematics saw a need to abolish causality in order to allow its own irrational interpretation to become believable. How can one live without causality? This would mean that all actions are without consequences. Drop a boulder on your foot and find out if causality is real. Try to see how far you can go in the world by disregarding causality.

Irrational mathematics is failing in a psychological interpretation of reality. Psychology is based on the existence of a conscious psyche with many levels of which only the upper level is outwardly conscious. It is our orientation in the world by judgment of value, which is the same as reason. This orientation is our mind which guides us through life. For example, gestalt psychology has proven that the human mind inadvertently organizes and connects random parts to make sense. Unconnected dots and lines are imagined as figures with meaning. Even noise can be organized into patterns. Ultimately, this means our minds organize fragments into meaningful gestalt.

Organization and disorganization are the recurring themes of the universe. This was mentioned by many others, notably by the Greek philosopher Heraclitus of Ephesus. The Rorschach ink blots, used in testing the imagery of mentally confused persons, are another example. The test helps tell psychiatrists how a patient's imagination creates fantasies. These are additional examples of how our imagination organizes fragments into meaningful gestalt. For organization to happen, suggestibility is required. Where did suggestibility come from? It was always there, we are only now beginning to understand what role it plays in the phenomenon of reality. We can assume it is the principle inspiration to the creation of value and meaning, which is all that reality ultimately means.

Conservatives are skeptical. Unless something much better can replace the old and dependable model, why confuse people with a half-baked, incomplete model of reality? This would be correct thinking, but in the last fifty years we have failed to discover anything that was not already known. At some point we must sort out and organize what we know into a new model. This point I think is now!

I assume that in the distant future radically new discoveries will be made, but they may not affect the already-established idea that reality is a value-creating process. If one is waiting for absolute certainty, that will never happen. The future will always remain unknown.

Without doubt, empirical science has a commitment to experiment. It therefore stands opposed to speculative and transcendental explanations, which in this case mean suggestion. But suggestion and suggestibility are experimentally verifiable and are therefore real. Admittedly, they are an inconvenience to mechanical thinking, because it is impossible to deny the presence of suggestibility and organization in the world. Although on the material level suggestion has only a subtle influence, to expect suggestion to have *direct* effect on material objects is not a human mandate. The separation of levels and modes is necessary to prevent confusion.

If one knows that during a trance cells respond to suggestion, one must also know that the cells are made up of complex chemical structures of chains of molecules and atoms. This means that molecules and atoms are also suggestible, a point confirmed by nuclear physics, again on a different mode. This leads to the conclusion that all things made from wave energy are suggestible. This, of course, means everything in the universe is suggestible, because everything is organized from waves—a phenomenon that expresses itself in the variety of meaningful ideas that create worlds.

Scientists may want to say that this deduction is too farfetched, but no more so than their own mathematical theories that want to attach mechanical laws to everything. Let us switch to a detail. It has been proven that the brain functions by adjusting to increased demand. The brain adds more receptors and expands in awareness, but will reverse the expansion if unused. This implies a flexible consciousness suggestible to value.

Enough of science for now. Technology also deserves attention. Although the two interact in principle, they are not the same in details. Two of the strongest foundations of engineering are geometry and mathematics. They were long considered the sanctum of truth, and became the best practical method to design mechanical and electrical devices. After muscle power, kinetic and solar energy were developed, but chemical conversion of energy by combustion of coal, oil and uranium are now in common use, causing a serious problem with the pollution of nature. This is an indication that we are on the wrong track with any mechanical technology that needs external sources to power it.

Therefore, naturally existing forms of energy are preferred. They include water, wind, the sun and all living beings. Many scientists and engineers have speculated about the use of such sources of energy, yet a major problem is inconsistency of availability. What strikes me as interesting is the synchronisity in which solutions are found. It is as if it were arranged that ideas evolve in the collective unconscious and persons come forward when needed—sometimes before the ideas are accepted causing them to be ignored.

In ancient times, Herodotus would have explained the coordination of events by providence, and later Jung would refer to it as synchronisity of necessary events in psychology. Even the scientifically oriented philosopher Edmund Husserl, after long years of phenomenological reduction, arrived at the necessary explanation of "intentionality" in the world process. This is not much different from the idea of providence or higher guidance. So after two thousand years of reasoning about reality, we are left with a similar mysterious and uncertain guidance. Is there a guidance? It is unlike a human idea of any of the gods of the past.

Incidently, Husserl was the closest that a philosopher could be to a mechanical scientist. He used mathematics, physics and even psychology combined in a careful analysis and in the end was forced to reach the same conclusion as Hegel and the others before him. And I continue to further define this intentionality as suggestibility to value, innate in cosmic consciousness.

Certainly, mathematics is indispensable in our mechanized world, but it can be taken to an extreme. Quantification is fundamental for technology and for the organization of commerce and finance. It is important for all such things that can be standardized first and calculated next, but only as approximation, still subject to frequent readjustment to actual values in reality. But when the difference between actual and suggested value is too great, the financial market crashes and so will our natural system of ecology. It is a fact that tells us we must return to actual values in nature.

In theory, exact knowledge would be ideal, but this is impossible in a changing world. In praxis we find out that even standardized materials are not reliable and we need extra safety margins. Materials change in time and manufacturing defects are frequent. For this reason, statistical averages are more dependable in calculations, and data is often questionable.

It was inevitable that mathematics had to lose some of its luster. Although its structure is similar to nature in regard to levels, each level is a step up from the previous, and on the higher levels it becomes multidimensional. Mathematics is a method of quantification, a system into which real values have to be inserted for meaning. This is where it becomes uncertain. For instance, money is also an abstract concept that can be calculated to the last fraction of a penny, but the real value is something else. The price of things depends on demand, and all calculations come to nothing when the values of real objects falls rapidly. At this point the emotional character of reality asserts itself, telling us that the energy that drives the world is emotional. Financial markets reflect this fact.

Material compounds are more stable than organic, which makes them trustworthy for use in engineering. Since the material level is without design and structure, a new level on which ideas come to prominence has evolved. On this level, ideas organize chemical substance into meaningful designs, namely at the molecular level.

Because materials are passive, self-absorbed, autistic entities, technology is able to reprocess materials to construct machines. This we can do, because our intelligence is able to reorganize passive beings for our purposes without causing them to revolt. But many apparent chemical reactions are actually emotional responses, such as fire reacting to water with violent anger, or dynamite with explosive emotional force. The only difference between a mechanical model and an emotional one is that the lower levels are predictable, making them scientific. When chemicals can be joined in a synthesis, they may like each other. Or they can be mostly disinterested, tolerating or ignoring each other unless a catalyst is added to organize them into new materials, as in plastics. Often chemical combinations do not like each other and collide in violent reaction, called combustion. In general, the passive level also has emotions and fits into the system of levels and countless separations which keep modes apart by separation of value identity.

A pragmatist would say that as long as something works, that is all that matters and we should make money when we can. That may have worked when our planet was underpopulated, but now nature cannot recover from such misuse. If one understands the underlying reasons, one will not make mistakes with catastrophic consequences. This demands a sense of harmony, composition, and innate suggestibility that has feelings other than utility and dominance, to evolve culture and dramatic art and to make life an interesting experience. To think of cash value is shortsighted pragmatism.

The development of digital technology has resulted in methods of bypassing many unnecessary mechanical devices to simplify machinery for mass production. But even more interesting, microwave technology has resulted in improving the spread of entertainment. As such, it does not help create more material value, but makes life an interesting experience by giving people "quality time." The computer was initially considered a practical device for organizing information, with a useful profit-making side. Now it has become an instrument to extend memory and human intention, which is a reflection of the Universal Principle to experience meaning. This intent is expressed particularly in the spread of the worldwide Internet through the connection of computers. Utility has become a secondary concern. People are more interested in sharing experiences with other people, than in commercial or technical information. The information highway is in reality a way to share experiences—something that we humans have already done for a long time. In sleep, we dream of realistic experiences in which we participate in the affairs of other humans. One has to agree that the mind can create the experience of reality in dreams, without the need for physical manifestation. In the waking state, however, we live in a continuous physical manifest reality.

We are aware that values are labels which practically minded people like to attach to things and beings in order to classify and catalogue. But instead they have to make concession to a changing reality by switching to statistics and thereby greatly reducing the margin of error. With statistics they find that, in general, humans are predictable, but in detail individuals are still unpredictable. Nevertheless, people prefer labeling as a general guide, but should be prepared when change is required, to readjusted their judgements to facts.

The evolution of consciousness leads beyond the physical level, where objective methods are difficult to apply, and electronic technology has taken a turn to organize waves. It has created electronic phenomena, similar to the human brain, with one major flaw: The circuits are fixed, based on mechanical technology, while the living brain is able to change to different conditions because it is conscious. Intelligent and capricious individuals can outthink mechanical functions, and in thought, can transcend physical limits. With this insight, technology is still valid on the material level, but fails beyond on the psychological and emotional dimension. As an explanation of the principle of reality it is not just misleading, but far too narrow a perspective to explain nature.

Yet science deserves our admiration. Together with technology it has advanced our civilization in ways we dare not to admit. That does not allow anyone to think that artificial inventions are the way nature has evolved. It is ridiculous to claim that physics is near to formulating the Universal Principle as a Grand Unified Theory that can explain everything with one mathematical equation. This claim is so far from reality that we should ignore it, along with the Big Bang theory of creation, as unreal compared to the organization from galaxies to man. As intelligent beings we are responsible for what will happen if we disregard nature, namely a colossal catastrophe, and as species we should have still connections to the collective consciousness. Collective consciousness appears to guide us, but at a distance and indirectly, because we are meant to be free agents to experience meaning and accept responsibility for our actions and should not rely on others to help us.

7

The Structures of Reality

The greatest step in our understanding of the structure of reality was the breakthrough to the level of electromagnetic energy, radically changing our perception of the universe. Everything that exists has become a phenomenon created by invisible waves in frequencies that appear solid to us. This means reality is based on the synchronization of wave-frequencies which our mind interprets as relevant to the experience of the tangible world.

The breakthrough did not come as a surprise. It was preceded by philosophical ideas and discoveries from other directions that came together in the twentieth century as wave-phenomena without physical substance. Philosophers and mystics suspected that reality is a phenomenon of consciousness that created the world of illusion. Hegel called his philosophy the "phenomenology of spirit" because it was a consensus of idealism. The other direction came from discoveries in psychology. Mesmer first believed that electricity in frog legs (discovered by Galvani) meant that physical illness could be cured by magnets. But he learned that he didn't need magnets because people were cured by his presence alone. This indicates that health and illness are self-suggested. The third, and most convincing direction came from the discovery of electricity. The electron was explained by J.J. Thomson as consisting of invisible energy without substance. This was followed by many other electronic phenomena, including the discovery that light could be created through excitation of molecules to make metal wires glow. This was applied by Thomas Edison, inventor of the light bulb.

About 1925 such inventions convinced German physicists to give up on the physical concept of materiality and to develop a wave-theory based on the "quantum" discovered by Planck. With this step, the universe became a phenomenon of waves. Gravity and matter became unnecessary. Reality could be explained as the synchronization of waves, which at certain frequencies appear as solid.

If one connects all three concepts, reality becomes a phenomenon of conscious wave-energy organized into meaningful ideas by innate suggestibility to value.

The question of who suggests ideas that appear real is open to speculation. If they were the creation, of a god, why does the world evolve by trial and error? This indicates that innate unconscious suggestibility is more likely the cause. Yet, there are also indications that occasional interference from a higher source seem to correct the direction the process should take. This leaves the mysterious guidance unknown.

Let me go into more details to prove the existence of the stratified reality, separated by spaces between levels and the emotional energy to drive this phenomenon.

The question of whether the ultimate energy in the universe consists of waves or particles becomes irrelevant because it has been discovered that energy is emotional and suggestible. This means that emotion causes the waves and suggestiblity the organization of waves into meaning. This explains light as emotional excitement and that it has a positive and negative side. As positive emotion, it can increase in energy, and as negative, decrease. This implies that our sun need not consume its fixed energy but create more energy when needed. That contradicts all calculations of astrophysics about the duration of the universe based on thermodynamics. The new model is a psychodynamic phenomenon—a new vision of reality.

From this premise, the material model that is offered by science is obsolete. Through many centuries of philosophy and more recently, quantum physics, a consistent pattern has gradually emerged, showing meaningful organization from quantity to quality on the physical level. But beyond the physical, a similar organization of consciousness also takes place parallel to it. This organization of consciousness has evolved as further differentiation from quantity to quality as individuation, in refinement of emotional experiences toward compassion and sensitivity. We see this in the rise of culture and art, particularly the emotional excitement transformed in sound frequencies organized as music. A concert is an emotional experience on an immaterial level.

What I mean by structure is only partly visible—when it manifests itself physically and uses physical structures to express itself mechanically. Whether invisible or visible, it is the same structure, created by emotional excitement, that produces the phenomenal reality which assumes temporal organization of meaning inspired by the suggestion of value and then disorganizes when the excitement abates. From the physical pattern, one can make assumptions about what the invisible structure that organized nature might be like. As far as we can see, few beings in nature have sufficient self-consciousness to understand themselves. Man alone is beginning to speculate about the metaphysical process of organization and invisible structure that is beyond direct human perception.

In other words, the essence that inspired this phenomenon is outside the range of physical sense perception but not outside consciousness. It seems that this essence *is* cosmic consciousness, similar to our consciousness, which is also invisible but has a very real presence as an unconscious organization of nature. Without consciousness, there is no experience of reality. It is the essence of reality, both as unconscious or conscious orientation.

We know that the physical world has many levels, that consciousness also has many levels and that there are countless separations of values that we are not normally conscious of. We know that the purpose of the countless separations is to keep values apart to prevent confusion and that it took millions of years for the manifest world to organize. There is an underlying structure of values that gradually grew out of the unformed quantity to organize into separations of quality that have distinct value. In spite of the fact that all is a phenomenon of waves, it is manifested as solid, and is certainly not illusions. Nature has a structure that the suggestion of value and meaning has created; it is only partly visible, but its existence can be proven in many ways.

A general fact that seems to reconfirm itself is that the alternation of separation and organization are natural trends reflected in cell division. But this process is not without meaning. The apparent reason is to keep life interesting because when interest in life decreases, the suggestion of value weakens and disorganization follows. The decline begins when the immune system loses its ability to resist invasions and dysfunction actually causes death.

One should be aware that understanding reality is more than a philosophical exercise in metaphysics. In fact, the physical universe has manifested itself gradually through the ability of waves to assume frequencies that appear impenetrable and solid. Within this range of frequencies, waves have synchronized as meaningful, stable objects. For that reason, Hindu philosophy is not doing us much good by calling the world "maya" or illusion. Nor did David Hume or Bishop Berkeley perceive reality correctly, by saying all things exist only in our minds. On the contrary, manifest beings have values of their own, as do we, in a universe of many levels and ideas. But only consciousness is real, because manifestations are only temporal phenomena.

Apparently physical ideas organize from waves that are manifested as real, and become solid matter to serve as the structure for consciousness. In a general sense, the physical structure is the infrastructure for the psyche, a temporary frame of reference, used as a house for the invisible soul to suggest value and experience meaning for a while. From the many indications already mentioned, only the physical structures disintegrate. The design of the organization is transferred by seminal propagation to be reorganized into a new physical structure, implying that some memory continues in the genes, even in an improved design, thus indicating evolution. This also would mean that over time a complex fabric has developed that suggests solidness and realism to life, but is secondary to the experience of meaning.

We can only speculate about the extent of the universe and how many levels or frequencies might exist. We can understand the underlying intention, namely to experience meaning, when we consider the meaning of ideas and what the purpose of the whole might be. Ideas have meaning and nature has a structure. Together, they reveal a pattern that, in principle, follows one single intent—to be something, anything and everything that is more than nothing. This is reflected in the pattern of reality.

Practical wave technology can be sued to create meaningful patterns that allow the development of ideas, such as programs on television and video. A particularly interesting example is the creation of holographs, three-dimensional and lifelike, produced by loops in the organization of waves. But we never can give them a living consciousness or physical solidness, because living manifestations take the slow route of evolution.

Since I consider the physical side of the world as secondary infrastructure, it is consciousness that has made the greater progress, the body has changed little. The only explanation can be that change is more interesting than a stable existence. Perhaps the unconscious memory remembers many life experiences, but those memories can be a hindrance when the conscious mind confronts a new experience, so are kept in the unconscious. Perhaps there are moments of recollection—fleeting moments out of the past, including extrasensory perception, that occasionally become conscious by suggesting ideas. But we do not want to know the past or the future because every life is a new experience.

For technology and mechanization, the psychological part is set aside and comes into recall only when stable materials react out of style. For example, the solid ground under foot displays wave-like vibrations that transfer themselves through materials as earthquakes. When we experience hurricanes, electrical storms and tornados, we are reminded that materials are not dead. For a long time it would have been considered illogical to imply that consciousness is present in materials and electrical energy, but this has changed. Physicists have discovered that subatomic energy behaves as if it were observed, implying consciousness. While this may not change much for technology, it is important for a philosophy of nature. One can now explain the so-called mechanical laws of nature as the autistic behavior of conscious entities of lower awareness, which are oblivious to our manipulations.

The diversity that exists in nature is a linear progression of transformation from quantity to quality. The system of quantification used by science fails to consider quality, value and meaning, therefore, it has nothing in common with the real world.

I can understand this position, however, from a perspective of realism. For countless centuries, all sorts of magicians and prophets explained the world as spiritual, and philosophers, such as Hegel, claimed that ideas create everything. When all these claims achieved no results, realists started to experiment with material substances and gradually developed practical results. Early man discovered that fire could transform substances. Boiling and melting of substances started chemistry and then metals were made. Man is the only species that uses fire to improve the environment, and later electrical and nuclear energy to change nature to his own advantage. One can understand that scientists wanted nothing to do with magic, religion or philosophy. The entire field of the occult was considered useless speculation. Because of skeptical attitudes in academic circles, realism swept away metaphysical ideas. More open-minded persons still considered the possibility of extrasensory experiences and the population at large remained religiously minded. Even some academically trained persons began to think that there might be another side to reality beside physical proof. In fact, the originators of quantum physics, Planck, Schreodinger and Heisenberg, wrote that life also had a spiritual dimension.

Let us forget past mistakes: One can be misled by preconceived expectations and easily misinterpret experimental data. Forty years ago when physicists were starting to develop nuclear energy, some thought that world was created by fusion of atoms. A majority of physicists still probably accept that idea. Physicists became convinced that atoms could be fused together, just as many new substances, such as plastics, had been artificially created through chemistry. But that may be impossible because the subatomic level is strongly disposed against change and on a different mode. In fact, in fifty years of experiments with nuclear energy, only fission has been achieved. This happened because uranium 235 is a naturally decaying isotope, but fusion is not based on decay. Fusion is upward mobility that belongs to organization, and remains difficult and elusive.

Because extremely high temperatures would be required for fusion to occur, even the temperatures of stars are considered too low. Physicists think that the complex atoms were all created in a grand cosmic explosion. The Big Bang theory is based on the belief that the fusion of atoms happened in the first second of the primal explosion, when the necessary temperature was reached and the universe was then created. Obviously, chemical synthesis is artificial creation and has little in common with natural organization.

Complex atoms are necessary to explain the vast variety of things that make up the material world, but this does not mean that nature had to create these atoms by forced fusion in the first split-second of the cosmic explosion. There can be other explanations, such as organization. The assumption of fusion is based on the belief that the grand explosions of the entire mass of matter in the universe created the necessary heat to cause the fusion of atoms.

Philosophy reacted to the wave concept, and a new type of scientifically oriented phenomenology, led by Husserl, became the major subject of investigation. He undertook what was the most comprehensive reevaluation of knowledge in our time, combining physics and psychology, matter and idea, and after decades arrived at the conclusion that reality is an "intentional process."

We know that conscious orientation has slowly evolved until it reached self-understanding in mankind. Organization by ideas, as proposed a century earlier by Hegel's idealistic phenomenology of spirit, has been confirmed in principle. This conclusion indirectly confirms innate suggestibility to value and meaning as the cause of organization of nature.

Because philosophy alone is not enough proof for skeptics, scientific proof is sought through experiments in human perception. Such scientific proof was rendered by experiments in human perception. Because perception is based on consciousness, I consider the experiments performed by gestalt psychology as proof that consciousness is trying to organize what it perceives into patterns of meaning.

The experiment consisted of having people look at loosely strewn fragments of lines. The human mind inadvertently organized what were random fragments into patterns of meaning. The patterns assumed a meaningful gestalt in the mind of observers. This, in my understanding, proves innate suggestibility to value in conscious orientation. Obviously imagination, a fact of consciousness, is involved. This also is confirmed through use of the Rorschach ink blot tests used in psychiatry.

I would call this experimental proof because physical objects are involved. That the conscious mind organizes perception into meaning is very obvious because we constantly think about what our perceptions mean. This includes the arrangement of thought into meaningful patterns which provide us with flexibility, a characteristic of conscious beings.

Externally, individuals must be assertive and protective, even forceful, to survive in the demanding world. Yet, within the shell of each individual's self, the unconscious identity is open to suggestion that can influence the organization of the physical body. And internal suggestions can also influence the organization of the unconscious mind such as its self-image of worth. In the external world where persons freely interact, one has to stand up for oneself and, if needed use force, because persuasion is difficult when reason is absent. Just to live we must compete with other organic beings. Since all things resist being changed by others, change becomes the challenge of existence.

Depending on the exposure, positive persons remain unaffected by life's tribulations. On the other hand, negatively disposed persons develop physical or mental problems, which are indirectly self-suggested by one's own feelings of inferiority. Therefore, the protection of personal values is essential to good health. This implies that one should avoid values that contradict oneself. But fixed individuals could ignore change and then would never progress or improve.

Most reasonable persons believe that one can sustain some abuse, physical or mental, and as long as their balance is not irreversibly disturbed, they will recover. This is also one of the most controversial subjects where disagreement is wide spread, and can be a test for the truth of the validity of the Universal Principle of suggestibility to value.

For example, with organic illnesses that affect internal processes ranging from gastric problems to allergies and including arthritis, paralysis and cancer—the medical profession is split on whether the causes are emotional or physical. The majority wants to cure the problems with drugs or operations to eliminate what they think is the physical cause. But if this results in the escalation of drug use and operations without a cure, then the cause is emotional. Obviously, one can mistake the symptoms for the cause, particularly if all physical causes indirectly are emotional, because in principle even materials are suggestible.

The other approach is psychological. One can call all physical problems emotional. I assume this to be correct because I want to arrive at a single cause for everything. Freud cured progressive paralysis by making his patient understand that the problem was in the patient's mind. The cure of paralysis worked faster with hypnotherapy, but with arthritis, allergies and cancer success is questionable. What was the reason? If there are also other causes, then the single cause for everything is wrong and the entire philosophy presented here is a failure. For a while I gave up, but the presence of separate levels in the unconscious saved the idea for me. Just as fire-walking is also not possible through hypnosis, this means a deeper emotional mode is required to cure disease. It would occur on another level, almost as in psychokinesis, and it would mean that levels have different modes, only a fearless, selfless disregard of danger, like mystics have, to conquer emotions, including the fear of death by assuming that consciousness is immortal.

I was encouraged in my thinking because occasionally cases are reported where creative imagining has resulted in the remission of cancer. It came to my mind that ultimately self-suggestion is how the unconscious organization can be reached. If a person cannot believe he is the cause for his own suffering, he cannot understand that the body has its own ideas. Why should anyone believe it? The controversy continues as long as proof is absent and a physical cause might be found. Yet, proof has finally arrived. Microbiology has discovered that arthritis occurs when white blood cells attack the cartilage of joints and affect the bones, causing great pain and suffering. But, the cells are suggestible, which means that arthritis is therefore self-inflicted, as if one wishes to suffer. The same situation applies to other suffering that occurs when the immune system is disposed to make life unpleasant, but not to end it. This is not as critical, however, as when white blood cells allow cancer cells to grow wild and disrupt organic functions. This is terminal to the organization. This very same situation happens with mental illness when the mind becomes disillusioned with life. Gradually, disorganization results and can lead to organic dysfunction and death.

The question is, can developing an internal sickness be a death wish that an unhappy person has? Is this a wish based on unconscious discontent with his role in life? If so, why does he chose long painful suffering instead of a fast ending? Such questions lead to the psychology of self-punishment by guilt. Many will not consider this possibility because they fail to understand that the organization of cells is controlled by the unconscious level.

Such speculation fits into the system of separate levels on which the protection of identity and value are based. Among many psychologists, Alfred Adler believes that most persons have an ideal, a self-image, that is usually set too high to reach. This may explain why they are unhappy with their existence, and unconsciously cause their own suffering because they have contempt for themselves as failures. This fits the pattern of the competitive, discontented individual who needs a physical handicap, yet consciously rejects this explanation. In fact, impassion and resoluteness are typical profiles of the angry person who has turned against himself. The profile of the healthy person is to forgive himself and others, to have tolerance with the imperfections of the world, to forget the past and to think ahead toward an improved future. In fact, typical of mental illness is the loss of hope for a better future.

From a neutral perspective, one can see this objectively. If the world were perfect, life would be boring. One needs causes and problems, and as long as the physical and emotional balance is regainable, one can recover. There is a clear difference between the emotional self and the external world as they interact between other beings and things. The internal person is emotional and protective, but the outer world is challenging and competitive.

The difference between a living person and the digital system of a computer is the same as the difference between a mechanical universe and a suggestible universe—namely organization by meaning. In the mechanical system, the emotional essence of consciousness, which is suggestible to value, is missing. Such systems cannot perceive changing situations, because consciousness is excluded. The problem is, if one allows conscious orientation to have a role, then reality becomes a subjective, emotional system, and the entire structure of the mechanical model becomes obsolete.

Among experimental scientists there are several who were interested in consciousness, but were degraded to a lower status, such as psychologists and parapsychologists that are classed as "soft" science. One could reverse this classification. The hard sciences became fixed by mechanical methods and the material dogma blocked acceptance of other modes of being. In fact, there are many modes with different values. Evolution is a value-creating process that works as differentiation, by separating and reorganizing of ideas that suggest values. It is the evolution of ideas, of which we are only seeing the physical mode.

John Lilly investigated telepathy in dolphins—their ability to anticipate human thought.[12] Later, as he tried to understand mystical experiences, he turned to studying the effects of frustration and physical deprivation on the mind. He wrote that during a meditation session he had an illumination in which he understood symbolically that consciousness is a stream of experiences like data from countless computers, all flowing together to one information center, which is the absolute, universal mind. His vision provides a metaphor of a man from our electronic age. He, too, comes up with the idea of a collective unconscious, reminding me of Jung and Hegel, who called this universal mind the Absolute, the World Spirit that becomes conscious in mankind. Many similar conclusions have been projected: the Will to Exist, Elan Vital and much earlier, Kundalini, or Chee. In theology it was the divine spirit of God, always with intent and discrimination, yet still unpredictable. Reasoning humans have never believed that this unknown essence could be a blind physical energy in a meaningless universe. But objective science refuses to make such conclusions about a meaning, thereby implying that the universe is a meaningless process—the worst of all possible scenarios.

Philosophy and science again came together in this century when quantum theory discovered traces of consciousness in electromagnetic energy and accepted uncertainty in detail. The majority of scientists, however, reject idealism as wishful thinking and think that quantum theory is unprovable.

I want to distance myself from all past ideas, because so much has recently changed; all quantification is ignoring quality which is that issue.

Physics has split the material atom and discovered invisible, immaterial waves, called electromagnetic energy. This energy, in my opinion, is caused by the emotional excitement of suggestible quanta which are the basis of being. Suggestibility is not only a necessary precondition to the existence of a meaningful world, but it is the orientation in all beings for explaining meaningful actions, particular in the rise of human intelligence and reason.

The fact that the structure of nature is physically real makes it logical to assume that there are additional levels. Countless extrasensory experiences demand the existence of another level. Reality is designed with a flexibility of change in the experience of meaning, which seems to be the intent of the universe.

This is certainly a metaphor, but it shows a better model of reality that is more compatible with all known phenomena in human experience than is the present mechanical concept.

Although the wave nature and reference to individual energy fields point to quantum mechanics, this is where the difference begins. The energy is not mechanical at all, but suggestible and organized by value. It is a process of the expansion of consciousness from simple to complex awareness, where the emotions become controlled by reason. In evolutionary terms, reason is a product of the expansion of consciousness by experience in the physically demanding world. A necessary requirement is the development of a distinct identity that is stable enough to exist without a physical frame. On the next level there may be other references to experience meaning.

Such a conclusion makes life on the physical level a test as to whether an incarnate being can exist without attachment to physical needs, to possessions and to the pleasures of the flesh. This deduction is not exactly based on moral and ethical concerns, but is also not devoid of them. It is based on the expectation of a higher experience of meaning than the physical level can provide.

This is certainly speculation, but not without references to other levels in the structure of reality. In principle, all religions consider life a test for a more meaningful experience in the next world, but they have a narrow universe of moral values as a reference. The wish for more knowledge might be enough to reach the spiritual level. In fact, the hope for a better life could be the cause for the creation of a more refined world to follow this cruel world.

From another perspective, the structure of the physical world is basically a parallel to the structure of the unseen psychic world that uses physical frames as temporary support to organize and establish many identities of value and meaning. The physical frames are only temporary manifestations because temporal structures disorganize when the organic body dysfunctions and the soul departs.

From a physical perspective, our body is a structure in which the mind lives. This structure can be seen as physical commitment. As Max Wertheimer said, "Gestalt is destiny." Some religions think the body is the prison of the soul and death is liberation. This would imply that structures are our fate, but mentally we are free to improvise from what we have inherited, to play roles within our given potential. This perspective is confirmed by history. We, as individuals, contribute to the larger theater of history and the outcome of the play remains uncertain because we act above mechanical predictability. It is above reason because we are also guided by suggestibility from the unconscious, and are meant to use our conscious judgements of value, which change with new situations. This requires an open mind and intelligence for guidance through the complex theater of values.

I know this sweeping generalization of a single principle needs more proof which can only be arrived at by comparing known to unknown patterns. This is exactly where skepticism can make demands, because the details are complex and the changing values are infinite. We know that the physical structure is solid. One cannot penetrate other levels in physical form, at least not while we are alive. But our reasoning mind can understand what other levels might be like. We understand something of the molecular mode from chemistry, the atomic mode from physics, and the wave mode from electronics, but only as response to our probing. The inner self, the identity of beings in other modes can only be sensed by emotional channels. We all posses basic innate suggestibility, but only a few individuals have developed their psychic talents. Regardless, general ideas from the unconscious are still more revealing than insisting on exact knowledge of changing details.

In fact, as long as the details are uncertain we still have something left to learn and the future is open. The idea of an expanding universe is more than a physical fact, it is a psychological incentive to life. In the process of the differentiation of values it is obvious that there is estrangement between species and individuals. If one is tolerant of the diversity of values, the world is an interesting experience. One can concentrate on a few cases of evil, and thereby overlook the fact that a majority is quite content with the world. Will the world ever improve? Unlikely. It seems intended to continue as it always has done because there are always new, emotionally impulsive beings born. Later they become tolerant and wise and know as they leave the world that little has changed. Each new generation must learn from its own experiences.

This leads to the logical conclusion that life, on the physical level, is intended to be a learning experience. This perhaps means that the world does not change, only that some of the people passing through it have expanded their consciousness, and in the process have acquire an individual identity which they might need on the next level. Please note, learning from experience does not imply that at the end of life there will be a judgement of deeds, punishment of sins or karma. It only means that one unconsciously seeks the level where one feels comfortable, as in evolution, by suggestible orientation to what one personally feels has value.

The structure of reality repeats itself in principle, is revealed by the psyche of the universe and is indirectly reflected in the human psyche. Although details can be out of line, conscious beings can adjust and correct them. In a mathematical system, a single mistake makes the entire effort useless, but with living beings, mistakes are forgiven. This means open-minded systems are forgiving, but fixed systems are both unforgiving and terminal.

If in observing structures we compare the experience of conscious beings we see that the only connection running through the entire structure is emotional. Values on all levels are organized differently. Only the external structure of levels repeats itself like a pyramid. Each level supports the next higher in complexity, to be stood on and to be exploited. Therefore, we all are part of the larger structure of the general value-creating process. If one thinks of the meaning of life as a trial, then a narrow religious interpretation can be implied. If one thinks of a world of physical needs and pleasures, and that beyond the physical experience is nothing, then materialism is a terminal concept because it denies the survival of consciousness. This convinces people that to strive for money and pleasure is all there is. There can be many other concepts of meaning, but whether the unconscious actually believes such opinions may be the final test. That means one cannot believe what people say, but how they live, determines whether they are idealists or materialists. In fact, their unconscious controls the physical organization and mental interests as expressed in the whole gestalt.

On the biological level, growth happens when cells divide and organize, and when a frame of hardened cells serves as structure for other soft cells to rise up on them. The hardened cells, which we call bones, provide a mechanical system of levers—legs, arms and hands—that are moved by the contraction of soft cells, called muscles, that provide mobility and action. But the important requirement is conscious orientation as suggestibility of when and what to move and why. The arrangement of levers can be seen as mechanical, but the cells have organized themselves from suggested ideas into a meaningful structure. Intelligent orientation is the issue.

Many mechanical elements are incorporated in the design of the structure. It is also true that chemicals are converted into energy from food, but this still is only the design of the infrastructure which should have a purpose. The purpose of the entire process is to experience meaning. The idea for the design for a mobile and self-sufficient being explains the creation of self-conscious beings able to know what they are doing, with exterior orientation. This generally describes humans as self-conscious beings partly controlling their unconscious impulses.

What we know is the exterior expression, the infrastructure of the world to which we can add the other part—the essence of being that is nonphysical, or metaphysical, which can be comprehended only by speculation. Speculation is going where science fears to tread and uncertainty begins. When physics arrived at electromagnetic energy, from the physical side, by measuring and calculating, scientists were thinking of a stable nature, of a Newtonian universe, that could be predicted. I never trust calculation. Despite a tenfold margin of security in mechanical things, they still can break down from a syndrome called "fatigue"—weariness from bodily and mental exertion—as if they had feelings. But I am talking about fatigue in metals, which implies that material can become tired, as if the energy that holds materials together is emotional, and loses interest. The same energy which in emotional excitement lights up the darkness of space can become disillusioned and physical structures fall apart. Indeed, I jumped from the strengths of metals to cosmology, to the "light of emotions" as if it were one and the same as electrical energy. I concluded that in addition to emotions, the electromagnetic energy has elementary consciousness, which is suggestible to value. But, if there is only one essence, all makes sense.

Let us take an example from calculations made by physicists about the mass of nuclear energy in the sun. According to them, the sun should have burned out long ego, but instead it keeps going.

This means the energy is adjusting to needs, is suggestible and emotional. In fact, the essence of nature is unpredictable. It can become emotionally tired or fatigued, or it can also become excited and develop more energy as it is driven by an idea that suggests value to become motivated to act. This is a comparison to human emotions, in principle only we humans can also generate emotional excitement. Our electrical field is organized by a long evolution of ideas that can raise the energy level in a civilized society.

Many names have been proposed to describe the nature of this low emotional energy. For example, Kundalini, rising from sexual energy in the spine to evolve as intelligence in the head, may be near to the reality of human evolution. As one grows up, the body and mind organize and orientate themselves. When one reaches puberty and vital energy becomes sexually oriented. But a person also can have intellectual interests and will evolve by expansion of the mental horizon early in life. Individuals are inspired differently. As people age there is a wider separation from physically predisposed to intellectually oriented values. Some people will lean toward science, others to intuitive fields of philosophy and religion.

This is a controversial subject. I can see no preference given by nature for moral or religious beliefs other than for self-satisfaction and tolerance toward all beings. I know the argument that the scientifically oriented raise against the continuation of consciousness without a body. Consciousness ends with the body because it is only a by-product of electrochemical processes and will end when the body disorganizes.

We have discovered that our planet, the Earth, has transformed and altered its orbit several times and that it has reversed polarity, causing radical changes for the beings that evolved on its surface. As radical changes in climate occurred, less vicious animals—the mammals—replaced the reptiles. There is a lesson in this for us humans: If through our actions we cannot keep the system of nature in balance, nature will have to replace us with beings that can control their violent impulses. One can assume this happened to the once dominant reptiles. This idea is only speculation, but has parallels in human life. Reasonable persons do advance past impulsive individuals who get themselves into conflicts.

The process of organization, by which nature indirectly influences evolution, requires something like a collective form of mass-suggestion to effect major changes in organization. Therefore, a practical way to understand the arrangement of nature is to start with collective ideas that organize cells and structures by adding and securing layers upon layers in the organic and mental structure of reality.

It so happens that we only have a limited influence on the organization of the body. The unconscious mind keeps only whatever fits in its pattern of organizations. On the dream level, the deeper unconscious mind uses symbols that resemble physical structures, such as houses, animals and even mechanical objects such as automobiles and trains as metaphors in the special language of dreaming. This implies that the consciousness uses known structures to convey ideas.

This information is only useful if one pays attention to it and learns to interpret dreams. If one ignores dreams, one has closed a door to the unconscious level called the surreal. But dreaming continues and the unconscious communicates on a subconscious level regardless whether we pay attention.

There is agreement among professionals about the meaning of frequently occurring dream symbols such as the house. Interpretations indicate that the unconscious is comparing the human body to a house as a temporal structure in which the "self" dwells. This indicates that the body is to the unconscious mind only a temporal abode, which the spirit-soul must leave when the body becomes dysfunctional. This implies that the soul survives, meaning that the visible world is the place for conscious-soul entities to experience meanings in other manifestations.

This is certainly metaphysical speculation. Yet, continued in a linear direction, the structure of levels appears to be parallel to the unconscious structure of the mind. It is obvious that dreams are uncertain. Our knowledge of the psychic side is uncertain, but it is more realistic when compared to scientific materialism that denies consciousness, calls organization coincidental and reason misleading.

Some believe there is a higher power, which provides guidance from a distance. This guiding power may influence the theater of human affairs by changing players and indirectly orchestrate the outcome of events by making subconscious suggestions to human minds. Then, humans think the ideas were their own. This explanation may seem speculation, but some personalities, to whom such novel ideas have occurred, confirm that their ideas came from transcendental inspiration. Intuitive and transcendental insights are confirmed by inventors and other thinkers as flashes of illumination. From religious experiences to artistic inspiration, to writers and philosophers, even scientists acknowledge that important ideas came to them when they had almost given up. Nietzsche compared nature to a capricious woman: The more you want her, the less she wants you. But if she sees you give up, she might let you have what you so diligently pursued. That could be true of ideas and inspiration: When one is ready to give up is when an idea occurs. Is it because nature wants us to feel important, to make our own discoveries? Perhaps this is the reason why suggestions from the unconscious are left open to interpretation. It is to give us the freedom of choice. Yet freedom to have a will of one's own does not guarantee that the choice is a wise one. Suggestions must be judged for value—that is where intelligence helps.

The skeptics are correct. Ideas can be misleading, but skepticism can be paralyzing. There is nothing else left but good value judgement—mechanical methods are too simplistic. Even the religions are correct on a main point. We are exposed to good and evil suggestions from the gods, but it depends about which particular god are we talking about. The gods of many Asian people, such as Vishnu, are creators and destroyers simultaneously. When the pagan Europeans became Christians, they adopted the idea of a just God, but justice had to be deferred to a next life beyond earthly existence, making it an uncertain promise. Humans have many ideas about gods. Therefore, the philosophers are right to believe that God is an idea that people need to explain the world. God is a metaphor to understand the meaning of life, to fill a vacuum when nothing else is known.

With our present knowledge, the idea of God has become an observer who gave us intelligence and responsibility for the organic level of our planet. With our increased knowledge, we become free, at least in part, to make the best of life. What lies beyond death is unknown, and therefore is interesting because it implies that we have a future.

Humans speculate about the meaning of life and what follows. Most no longer feel like slaves, shaking in fear of hell fire, but feel challenged to act intelligently and responsibly with influence on the management of the environment.

As a cultural tradition, religion has had the useful social function of providing a spiritual dimension. Science, however, has not been able to provide a useful alternative, only negative answers.

Metaphysics suggests positive alternatives that make sense and correlate with reality. Current psychology and psychiatry indirectly confirm the existence of innate suggestibility in human experiences. It is under stress that transcendence is normally blocked, yet it can be temporarily lifted in exceptional situations. Nature uses transcendence only when dramatic experiences are involved, thereby causing humans to speculate about possible ways to develop transcendence.

Psychokinesis shows that in emotionally charged situations with persisting frustration normally impossible phenomena can happen. This implies that transcendence offers an escape from bad situations. The system of nature shows compassion for suffering, such as in the critical moment of a dream when the dreamer is saved by waking up and, once awake, thinks that what has occurred is only a dream. Perhaps, this reveals something about death—that one may wake up in another world? In real life suffering can become drawn out before one is released from the physical body, making escape desirable.

The structure of nature is designed in separate levels, compartments and individual entities, indicating that separation is the main intent underlying the structure of the universal psyche, and that it is meant to provide privacy if desired. The entire structure of nature is emotionally connected. If there is a common danger, beings will cooperate and support each other. Otherwise beings do not desire to get involved in their private experiences.

Normal people agree that individual privacy is desirable because one is a person with individual needs. The provision for privacy is reflected in the design of the structure of the known physical world, which is in principle like the design of the unconscious side. Normal people do not desire to transcend time unless it happens involuntarily. Spiritualism and the occult are intrusions into the unconscious—unless they come from the unconscious. But when transcendence, such as clairvoyance happens from within, the information is too general to be useful to avoid impending events from happening except to serve as emotional preparation for them. This makes for agreement with religious thought—that somewhere on a higher level, compassion exists. However, the concept of noninterference which provides freedom of choice and responsibility for our actions overrules the desire to help individuals so we must evolve on our own. This seems to explain the general absence of divine interference in human affairs and nature. The main goal seems to turn us into self-sufficient and intelligent persons who are meant to develop sound value judgements, to live by reason and to become autonomous beings.

The ancient idea of the immortality of the spirit and reincarnation are some of the ideas learned from extrasensory and transcendental experiences. Many cultures share a belief that the soul survives. A contribution to the understanding of the psyche of nature was also made by the American mythologist Joseph Campbell, who based his ideas on the heroic age and the self-sacrifice of the hero. This may be a part of the psyche of nature that sacrifices human lives. According to Johann W. Goethe, a German poet, man evolves from death to new life in the act of "die and become." But I think that the development in recent history has forced mankind to shift to a higher level and mode, so that heroism has become obsolete. Even self-sacrifice or any form of killing is a primitive impulse. The development of atomic bombs made war a mutual suicide of mankind. We are being forced by technological progress to control our emotions and to advance to a higher level of reason—in line with the progress of civilized society.

The idea of the immortality of the souls has been reinvestigated by Raymond A. Moody, who collected information from the personal experiences of hundreds of people who were revived from death by medical intervention. According to their statements, they experienced the separation from their body and were floating above as a conscious, amorphous entity in an invisible field of energy. In our electronic age, this is conceivable as individual quantum field in a phenomenal universe of multiple frequencies and wave length. In addition to the scientific compatibility, the emotional side also requires consciousness. All else is relative including the experience of a bright light. Those experiencing this separation say that they were indirectly asked whether the experience of life was worth it, and whether they wanted to return to the physical level. As I read Moody's accounts, my own belief in the survival of consciousness was strengthened. In addition, the unconditional love that is emanated from the bright light implies that moral guilt is not an issue beyond death. Philosophically, this makes sense, because in a value-creating process opposites create each other. Without the suggestion of evil, goodness has no meaning. To bring it in line with a single emotion, unconditional love is the same as unconditional value. The bright light then is a higher-evolved version of the emotional excitement of the cosmic light that illuminates the worlds, creating and sustaining all experiences and meanings in the universe.

One might think that a philosophical attempt to understand nature in a total sense is an idea whose time has come. The support from science for the idea of a Universal Principle should come together at a time when it is most needed. It is needed to enlighten us about the progress of technology that now endanger nature's ability to survival that has taken so long to evolve. Many events seem to have come together at the right time to prevent an ecological catastrophe of major consequences for the organic level of our planet.

A good reason mankind would benefit from understanding the principle of nature, is nature's own concern for survival. What use is it to nature to withhold secrets from us when science and technology are already capable of destroying the biosphere? If left unchallenged, they would unleash the emotional energy of the primal level, which would destroy the higher organized wave forms. Our ignorance is of greater danger to nature than keeping the knowledge of the purpose of reality a secret.

It may be a disappointment to physicists that the creative process as evolution of consciousness happens by internal organization. Conscious life is technically impossible to reproduce, but it is a wise precaution of nature to keep ambitious individuals from misusing and confusing nature. Nature separates everything in countless ways to prevent undue influence. For example, nature keeps the possibility of a nuclear "chain reaction" from spreading as a conflagration to all atoms in the universe.

Nature, in principle, is a protective system. Understanding persons would agree that nothing less than separations guarantee the safety of our world. We all should be aware that in the real world, many capricious entities exist, including humans. In my opinion, it is a wise idea that we are committed to life and that life is limited. Commitment and limits makes this a cruel world, but it is still the best of all possible worlds. Without commitment to action by suggestibility to value, nothing would ever happen in the universe, and without limits too much could happen. (Voltaire was wrong.)

Therefore, we must agree with such philosophers who assume that a natural balance exists to keep everything in a proper order. At least some conflict is necessary for creation. Suffering also plays a part, because values need opposites for meaning. In the end, the process is beneficial to all as meaningful experience. And if the transcendental seers of mankind are correct, for example as expressed in the *Bhagavad Gita* and the *Rig Veda*, that nobody dies because consciousness is immortal, then suffering is only an inspiration to action in the world. In the end, all is forgiven in the unconditional love of the value-creating process.

Let us accept uncertainty in detail. Let us accept that consciousness will not end when the physical structure dysfunctions. We already have sufficient evidence about the survival of consciousness. In addition to transcendental experiences, we also have physical proof about the structure of reality as a continuing value-creating process. If this is not enough, nothing will convince a negative person. One has to learn to live with opposites.

The suggestions which organize waves into particles and particles into atoms with distinct qualities as their identity, are positively or negatively disposed. They are positive when they are happy and organized into ideas of values. But when they become fatigued and disillusioned, they disorganize to retreat to a lower state. With this explanation, a simple emotional principle of the psyche of the universe is confirmed.

Modern psychology confirms these facts. If one feels happy and content, one will be physically and mentally healthy. If one is unhappy and discontented either physical or mental disorders will make life miserable. Swinging between mania and depression, we are normally keeping to the center as part of the emotional psyche of the universe. The system of reality, which is already proven by technology to be a grand phenomenon of waves, has created itself a structure of levels that are separated and secured against collapse. We must be prepared to accept that other structures also exist, that might already have evolved further than we have. Unawareness of other dimensions is quite in line with the protection of values and identity. In fact, psychological blindness is a protection against superior beings. We could coexist with other worlds and never know it, at least from sensory perception. But through extrasensory channels, reserved for exceptional use, their existence could be possible. Because we have no control over it, it is difficult to prove that other dimensions exist and without proof, it will remain speculation.

Obviously, the speculative mind can see more than can our senses can because they are tuned to a limited range of frequencies. Philosophy has pondered this question. Socrates compared us to beings that live in a dark cave, who see shadows from the light and speculate about them. In a final sense, life is a personal experience, interpreted by the mind of each individual. Most experiences are common, but the arrangement by value is individually unique. Regardless, the need to have meaning, to be something, is necessary to remain viable.

If the need to have meaning is the same for individuals as it is for the combined consciousness of the whole, and if the whole is the combined consciousness, it could then perhaps be something like God. Then God is the sum of all values, the cause of the value-creating process. This also means we are all part of God. He should have compassion with the likes of us, after we played our roles—good and not so good. We should forgive ourselves and others, because we are only actors in a grand theater. We are controlled by situations. We have a body assigned at birth we did not choose, and we have unequal physical and mental endowments. How then can one be punished eternally for our sins, if inequality is intended to keep the value-creating process going?

Philosophers of reason, such as Spinoza, Schelling and Hegel placed God into the world, as "reason," as the spirit inside everything, as the world-soul to which we all belong. With this shift in thinking God becomes like the other beings in the world, evolving to self-consciousness through experience—except that the process exists for a long time, perhaps eternally.

Heraclitus assumed the energy in the universe is emotional. We have rediscovered this concept from observing nature in greater detail and concluded that the human psyche is related to the psyche of the universe, implying the cosmos is a psychodynamic process.

In a universe organized by innate suggestibility, value is the recurring motive, but different values are the experience of each species. Values separate and consciousness indirectly connects everything. But there is a higher level of consciousness that transcends the separations of value by remaining neutral and agreeable, by respecting differences, and by valuing all things equally, that creates peace. This fairness opens doors, at least in many beings, depending on response. As was mentioned before, one must be selfless.

If one is a conscious soul existing in the void, in a state of limbo, in total nothingness, then the idea of experiencing something of value and meaning is exciting. Because we are *already* something, this idea no longer causes us to burn with the same emotional excitement that causes galaxies to ignite and worlds to organize. A suggestion must be powerful to have meaning. Just as many things happen to us in frustration, emotion can create things, and massive emotions can create massive things, such as galaxies and worlds. It must be assumed that massive emotions have given birth to the universe of stars and planets and organized worlds like ours.

Because suggestions are only temporal organizations from waves, they can disperse when the suggestion fails to convince. This fits into the scheme of things that exist in reality. It also explains why organized entities lose meaning, why they fade out and disorganize, and why they may desire to become reorganized again as different experiences, new and unburdened by old memories, with a fresh convictions to experience value in a manifest structure. This explains change as the only constant, because in time, all values lose worth, as one begin to take the value of life for granted, and a new beginning makes sense. Whether this means reincarnation or any other form of experience is unknowable, and may depend on the values and aspirations of the individual.

In human history, several reversals of the entire structure of values have occurred. Nietzsche, in particular, thought that a "transvaluation of all values" was necessary in our time and had one version in mind—a return to heroic values away from commercial materialism, which failed. A similar fate happened to the Marxist transvaluation which tried to socializes materialism. These transvaluations failed because they were forced on people. Values change from within, in their own slow speed.

Abnormal psychology is where suggestibility to values proves itself.

The emotional reaction displayed by individuals is essentially the same as that of the psyche of nature in general.

In principle, this emotional behavior can also be observed among stars. For example, the feeling of disillusionment may be the cause of a supernova in a star. In its last effort before it burns out, the star has a short spurt of assertion of value before dying. A black hole in a galaxy may be a mental depression. This is similar to deep hopelessness that some people feel, resulting in total disintegration. Cancer in humans may be an example of that hopelessness that manifests itself on the level of cells when a negative trend has spread in an organism. One could challenge this assumption, saying that radiation and chemotherapy have prolonged many lives. Indeed, but for a different reason. These treatments shock the body out of its negative trend and makes it fight back. Thus, it indirectly turns a person into a positive mode, which in turn awakens the cells and atoms to return to health. This is like electroshock therapy used on mental patients. It is also similar to homeopathy which challenges the immune system to fight back.

When the atom bomb was developed in Chicago, about 1943, scientists were surprised that refined uranium 235, an isotope, would not become a critical mass. Instead it released radiation outwardly. Eventually someone thought to enclose the decaying uranium with pure carbon bricks, thus preventing the radiation from dispersing and turning it back on itself, thereby heating the uranium pile until it reached a critical mass that could be triggered to either explode as a bomb or reach such temperatures as to cause a meltdown.

What has nuclear fission in common with human emotions and psychology? If one thinks of many negative and discontented humans placed together in a narrow enclosure, they would turn on themselves and become a critical mass, ready to explode. This is a way to explain nuclear fission as madness of energy. There are many examples of discontented persons, who have failed to release their anger, frequently keeping it inside until it turned into a critical mass. Suddenly, they committed insane acts of violence on imagined enemies. This happens when usually calm, reserved persons suddenly go berserk.

Happiness and suffering and all other opposites in values are essential to create the experience of reality. The majority of beings think it is far better to be "something" than nothing at all. This explains why we find new challenges in the continuing imperfection of nature, in a universe that remains forever incomplete and thereby infinite.

8

Human Affairs

WITH OUR PRESENT KNOWLEDGE of the universe as a psychodynamic process we understand the meaning of many events to be emotional and to involve values. It is comforting to know that the psyche of man is related to the psyche of nature, which is only a part of a much larger psychological universe. We understand that the problems that face mankind in general, and the individual in particular, are connected to the emotional psyche of the universe. It is one thing to understand something about the psyche in theory, but to use this knowledge in human relations is another. This is because all individuals are born with a strong self-assurance of worth and are unwilling to listen to what others want them to do and what to think.

When I give my opinion about the volatile subject of "human affairs," I know that I speak from a personal perspective. From a general overview of human history, one can notice periods, like levels, when values have changed, resulting in a mental transformation. People had different values to guide their thought processes. Let's consider early men who lived as hunter-gatherers, much like today's African bushmen. They lived as equals in matriarchal families. This began to change when more people lived together in villages. Tribes began to develop, levels of authority or rank were established, complete with a council and a chairman. Inevitably the council assumed executive power and the chairman made the decisions. With this natural development, the foundation was placed for a class system. As the world population increased, states were organized and people needed to defend themselves against invaders. So the military caste assumed control. This caste became the aristocracy from which a king was chosen as leader. This process happened independently around the world because it is rational. Yet, with it began the differentiation of people by class. If the difference were according to real merit, it would be fair. But the people who held office gave preference to their own children and relatives and this meant that merit was no longer important, just birth. In time, the class system became a feudal system because possession of land and wealth, not ability or merit was the difference among peoples.

Obviously this unfair system was often challenged. Frequent revolts and civil disobedience weakened the power of the aristocracy. When industrialization came, it brought a new class, the capitalists, into being. With their new wealth, the capitalists considered themselves equal to the upper caste. As commoners by birth, they did not receive the formal respect that the militaristic aristocracy commanded. In fact, most rational humans realized that everyone could become wealthy if they were lucky, or shrewd enough to exploit others. With this new thinking, there was a gradual change in the general population from a submissive outlook based on fear, to one of freedom and personal courage. The common people soon demanded their share of the land from the feudal lords and later from the wealth that came from industrialization. They organized into labor unions and became an opposition to management. We now cannot comprehend that people used to live in fear. Some were in fear of their physical freedom from their aristocratic masters who were in control of the military and the police. Others feared for their soul from suggested guilt by their religious middleman to God—the priestly caste—who claimed to be authorized to demand veneration. If only fear was the dominant factor of their lives, people would never have progressed. But fortunately they still had their individual self-worth that they held inside themselves. Each human being has a personal identity with the values that are necessary for them to survive emotionally. The first value is self-worth, next comes an innate loyalty to mother and family. As we look at the past, we see that by forcing their subjects into submission, the people identified with their masters and felt secure in belonging to a master a cut above themselves in value. Even if the suggestion of authority is not taken seriously at first, in time it will become believed, because the majority pretends it is so. Then, as people live under someone's influence, the idea of superiority becomes adopted as the normal way of life, and as such, becomes self-suggested. In fact, if one is born and raised into such a belief, one can take pride in belonging to a group or class that is loyal to an idea and its human representatives. Such a person may even become idolized and history makes him into a symbol of wisdom and courage.

In our time, the majority of people has grown up without apparent fear and blind loyalty and they are more likely to have the courage to stand up for their views and to think for themselves. They will think in ideas indirectly influenced by collective suggestions of conformity—which is socially correct because we need to have values. There never is a vacuum. Freely adopted values are accepted by the unconscious more easily than those forced on people. If many believe it, then it becomes a mass-suggestion. That is why mass-education, also can influence clear judgement. I am not just referring to religious education that produces loyal followers, but also to academic education that produces loyal believers. Scientific training contains just as much dogma as does religion. Scientific training is conformity based on respect of authority. It has discouraged many intelligent persons to think for themselves and instead they simply adopt mechanical methods without question.

From such comparisons one can see that the principle of suggesting value by authority is still the same, only the details have changed. Scientists usually are loyal to their indoctrination, for which they may have received the degree of doctor to prove that they have mastered a skill. Whether they have a wider understanding of the larger principle of reality is irrelevant to their specialized profession; they are secure in their values.

I feel that one has to trust reason and avoid mechanical methods, regardless of how widespread their influence is.

Man has also been called a reason-seeking animal, a step up from regular animals, which do not think much and prefer to live life as it comes. Thinking is the advantage which gives man a superior place in nature. Thinking is based on imagination, and if one has good judgement, then one can think intelligently by sorting out what is possible from the meaningless. An intelligent person knows that life is only a temporal experience. Thus life is a journey through a world in which one is exposed to many opportunities and temptations, a road strewn with obstacles. But without all these tribulations, life would be boring. Challenge is exactly what makes life interesting. Obviously, we are part of the emotional energy that permeates everything in the universe. Why then do many persons complain about life and would rather it were easy?

The reason is that one can become disillusioned when the obstacles seem insurmountable. One can think about problems in a negative way, and worry about what may lie ahead and ultimately fear death. This is a problem that animals do not seem to have, because they do not think and worry. This is a wise and good solution because animals live in a constant danger of being eaten so worry would paralyse them. Animals have taken death as inevitable and live without fear. Perhaps unconsciously they know that dying is only a transformation. This is what we humans also should do: Live fearlessly and think positively. Worry is destructive. Negative thinking can drive us insane if we allow it to preoccupy our consciousness. We can become sucked into a black hole like the galaxies in the heavens.

Negative expectation is the reason why we have seen an increase in mental illness and, in particular, widespread problems with teenaged crime. It is the reason that youths grow up to become adult criminals and must be either confined in prisons or mental institutions—separated from people with good intentions. Our current nihilistic interpretation of reality suggests that all there is to life is the physical world, in which cunning and egotism rules. With such ideas, one is tempted to take chances. If caught, one figures it was bad luck—next time one has to be smarter. Unfortunately, this idea has become widespread. There are many who think that life is simply what happens in the present, and that there is no other value or meaning to life. Materialistic nihilism is reaching to the highest level in human society. Intelligent persons, however, know that it is not worth the risk of their freedom or life for material objects. Many still take their chances in higher stakes, such as business ventures or politics.

If getting caught is the only reason why many refrain from crime, something is profoundly wrong with mankind. People lack ideals of personal worth and immortality. Immortality means more than passing ownership of material possessions because one cannot take them into the next level. Why not have a higher perspective? One can also try it out and expect that people are basically honest and loyal. If given respect and recognition, people will actually cooperate and help each other instead of exploiting and competing. The world is basically fair, and positive thinking persons create a positive response in others. Good-natured people are always willing to help others in need, as long as they are not exploited or considered a soft-minded, hopeless idealists who deserve to be relieved from all they own.

In fact, mutual cooperation has created the great achievements of mankind. Mutual cooperation is founded on trust in fairness, and has inspired loyalty in people that borders on self-sacrifice. When a person is obsessed with wealth and possessions, he is committed to maintain them and actually becomes possessed by possessions and is therefore limited in freedom. Materialism is a continuous problem of mankind, running through history as a block to intelligent living.

I have said already that the history of mankind can be reinterpreted by the Universal Principle and it will be different from current ideas. Certainly, the details are open to dispute, but the principle is laid down by the pattern that evolution has followed and we have to open our minds to recognize that it exists in reality.

Many would call me a naive idealist and mystic, but I think of myself as a realist. My understanding is based on cosmic facts, not mystical gravity and abstract mathematical equations. It is based on emotional energy whose excitement caused the frenzied spin of galaxies that heated itself as light, illuminating the heavens and organizing the worlds. The same light that we can see with large telescopes is physically real. Equally real is the separation into stars and planets. And so is the continued differentiation and organization into meaningful gestalt by ideas—that includes beings such as ourselves. The physical world reflects this pattern in the structure of reality.

The world's external orientation is our mind. It is a conscious level that has organized and civilized mankind along the same pattern by the congregation of humans into cooperatives and empires. These empires transformed and grew into democracies where individuals can evolve and differentiate by expansion of consciousness on an individual basis into self-understanding. An understanding of values is an essential part of a value-creating process, not a figment of my mind. The history of mankind is proof.

Let me return to the daily life of the majority. Most still believe that the majority is fair and honest. This belief, however, is attacked as stupidity from many sides. Many cave in to the relentless bombardment form the news media who report crime as noteworthy information and give it more exposure than it deserves. There is not enough attention give to what is good, fair and honest. The entertainment industry makes the criminal into a misunderstood hero. Today our children are exposed to what is called "merely entertainment." It comes from the distorted imagination of writers who would sell their souls for money. They actually believe that the masses want confirmation that vulgarity is acceptable. If one is bombarded constantly from an early age, then eventually the unconscious will believe that the world is a predatory place—particularly if the lifelike entertainment is portrayed as real life. Television and films have great influence on what people think. Scientific materialism has an influence on what writers think and imagine. In turn, the writers bring primitive emotions and materialism as popular entertainment, indirectly influencing the public. The idea of materialism, as truth, misleads the media, which in turn misleads the masses by suggesting a basically brutal and mercenary world.

How can this be acceptable to intelligent persons—people who know what is possible and what not—to allow fiction to sneak in the subconscious as fact? It happens mainly to children because they have not had a chance to develop reason. In a family where both parents must work to become rich, the children are left to their own interpretation of life and grow up confused. Somehow, the unconscious thinks in pictures and believes what can be seen is real. We have reached a stage in which creating impossible illusions has become an art, where reason is the only guide left, and what we see and hear may be totally false. That is where life experience comes in. Only individually acquired experiences can be trusted, and if one has acquired life experience through trial and error, life is over. Perhaps the world is meant to remain a learning experience where everyone is ignorant at birth and while passing through *can* acquire a personal identity and wisdom. "Can" is the proper word for it.

Philosophy has been called misleading by science, of course, the reverse is the reality. Starting with Galileo, mechanically minded science has been misleading mankind for three hundred years. It turned into the materialism that proclaims that the universe is a material process without meaning and suggests that life is final.

Not surprisingly, nihilism is spreading and materialism has undermined all beliefs in a purpose of the world. Nihilism physically and mentally prescribes a depressing scenario of doom for our planet. Only idealists expect that man can turn it around. The philosopher-poet Nietzsche foresaw the spread of nihilism through materialism as devaluation of life, but his ideas were elitist. He expected only a few humans to progress to a higher level of Overman by physical breeding. But it has not happened. If it occurs, it happens randomly and unpredictably and has nothing in common with genealogy or breeding.

Presently many could blame all evil on the materialistic, industrial progress. Why we are going totally in the wrong direction? Is it because profit and money have become more important than life? Nature also disregards life as it sees necessary. It could also mean that the value of life is declining because there are too many people in the world. When we were fewer and life was hard, life was dear. People now still can find challenges—they have moved to a less physical level, to the search of knowledge in uncharted dimensions. What more can eating and all physical pleasure give, if one can have them easily? The challenge is in what one does not yet know, beyond what is known. Man lives not by bread alone. We live by ideas—ideas that are more than a temporal perception in a journey through a short life—ideas contribute to the pool of knowledge. Few things are what they seem. There are many unseen factors that one misses at first sight. In fact, one sees with the mind. The mind organizes impressions by what makes sense, because the important qualities are invisible and have indirect consequences.

I have already explained that values only exist because separateness is maintained. Separation is necessary for anything to keep an identity. I expect that many separations exist in the world, many of them concern invisible things of which we only know from their consequences.

In nature, basic values are common and necessary for survival. Security, food, propagation and comfort are powerful suggestions we all share, but beyond that, values are individual. We all are born with different genes. We are born at different moments of astrological conjunction which affect our emotions, and with different socioeconomic backgrounds. These factors are part of the preconditions that cause diversity and dissonance and that inspire the need for reorganization and thereby perpetuate change. This point can be interpreted differently. As it turned out after many years to map a human genome, we have similar genes with monkeys, even flies. One gene that was isolated occurs in many other species in the same chemical order—and yet we are all different. Could it be that chemicals are only the carrier and the organization is actually suggested and carried as invisible electronic imprint?

One way is to think that nature is designed to evolve through creative reason. This view is long on history, and was also expressed by the philosopher Gottfried von Leibnitz as "the best of all possible worlds." Leibnitz lived in Hanover, Germany, during the age of feudalism when cultural and political ties with England were at their best. Newton argued with Leibnitz mostly about calculus, but Leibnitz also believed that the universe evolved from the organization of conscious monads, similar to souls. Newton was impressed by Galileo's method of quantification and the experimental method which he expanded to the idea of gravity as basic explanation for a grand mechanical cosmology, ignoring elliptical orbits of minor importance. Only in our time of space flight have we found out that there is nothing like gravity in interplanetary space. Once a spaceship gets above a narrow field that holds things to the ground, gravity does not exist and the mechanical universe is a mistake—the entire mechanical concept is unreal.

Another historical misunderstanding concerning moral issues also occurred to the French writer Voltaire, who took objection to Leibnitz's statement a generation earlier that this cruel world is "the best of all possible worlds." In *Candide* Voltaire criticized all the wrongs of the world. First, he ridiculed the aristocracy as conceited and simple in intelligence. Next, he attacked the church as inhumane in suppressing dissidents during the Spanish inquisition, and because it caused many wars such as the counter reformation that reduced central Europe to rubble. Voltaire's emphasis that religion has caused more evil than good depends on what one thinks about God. He also pointed to slavery, which at that time was still practiced, as inhuman, and criticized the use of conscription to coerce young men into military service, in the interest of feudal lords, just to be slaughtered. These conditions and many other evils which caused suffering to mankind made Voltaire ask, "How can this be called the best of all possible worlds?" Yes, the world is filled with suffering. But what would a perfect world be? Without opposites, nothing would have value.

In fact, most of the above mentioned evils have existed in history and have been repeated again and again. Life is not all evil. Among hundreds of good deeds, only one evil deed may occur, but because evil attracts attention, the good deeds are taken as normal. From a statistical perspective, we are basically goodwilled and loyal beings. It is just a few that give us a bad name. We cannot all be good, otherwise to be good would mean nothing. In this sense, Leibnitz is correct— it *is* the best of all possible worlds—but it could be better. In fact, Voltaire himself was no angel. He was pursued by creditors because of a frivolous life-style. This is just "not a perfect world," regardless what ever has been tried to change mankind. Changes have only been a different versions of the same principle. We have the recorded history to prove that transformations in values have always occurred, as further differentiation, like everything else in the evolution of consciousness.

To understand human affairs, the history of mankind and the parallel history of the ideas reveals the same general pattern which repeats itself everywhere. From the continuity of this pattern, the direction of the future of man can be presumed. It means that stratification and separation repeats itself as in mind and matter, in the unconscious as well as the conscious mind. In fact, the important changes that have advanced mankind and led to human civilization came from ideas. Ideas made the breakthrough of mental barriers in human affairs not just by levels, but as separated by species into distinct individual entities that have value for themselves. The separations are barriers that exist everywhere. Gradual change happens in distinct leaps.

Physical changes have not occurred in humans since the brain received the upper level addition, called the new brain, raising our consciousness above that of animals to conscious thought and since the addition of a voice box to express our thought in speech. These changes gave us the capacity to outthink other species and to communicate with other humans to convince them to join into larger organization with more power to take over other, less-united organizations. It is obvious that we communicate ideas in speech and gesture and suggest ideas directly or indirectly. In fact, we also can telepathically sense ideas. That is how speechless beings communicate in nature, but it is inferior in definition to clear speech which has developed to a written language.

With the addition of the two halves of the new brain, the two halves can exchange information, even talk to each other as ego and alter ego. With this change of design we jumped to a higher level of speculative reason. With this addition to our brain we begun to think about the world around us. Lesser conscious beings do not seem to care because the provision for thinking was not installed. Valid questions arise. Who installs functions? Who organizes the cells to a specific purpose? We cannot ask experimental science, because scientists refuse to speculate. They want proof. Speculation is not enough.

So let us think about why we have a brain with two sides. I would reason that the functions were installed because the improvement suggested value. But this cannot happen accidentally—certainly not as blind mechanical coincidence. Clearly, nature is a conscious process, and the mechanical improvements made by science and technology are artificial extensions of our intelligence. Without our senses, these mechanical improvements are useless. The fact that nature is organized in meaningful ways implies that suggestibility orientates itself by value. By this reasoning, innate suggestibility to value is the cause for all changes. If this is correct, it should confirm itself in all aspects of human affairs, as we shall find out through a glance at human history.

Human history

From the premise of innate suggestibility as the creative drive, the history of mankind becomes an emotional process that advances by suggestion of ideas and their implementation as infrastructure. Such an interpretation was already proposed by Hegel as the dialectical principle of the spirit of nature. In this principle, an idea is seen as a thesis, and in time every thesis creates enemies as antithesis. From the ensuing conflict, eventually a synthesis resolves the impass, which become the new thesis. This thesis establishes itself for while, but by imperfection creates a dissatisfied opposition which becomes the antithesis and the process repeats itself, thereby creating a chain of progress as evolution of ideas. This process was called the three-steps of history, as a continuous chain of mental progress to reach self-understanding in human spirit. But this idea was not applied to nature, where it would have explained evolution as "creative conflict." The dialectical principle was later misused in reverse order by Karl Marx and Friedrich Engels, to be reduced to a physical mechanism for the support of dialectical materialism, as theoretical basis for Communism. This example shows that ideas can have a colossal impact in the history of mankind, but materialism was basically a negative concept of a positive process. The dialectical principle was mechanized, as history has proven, because it was based on blind energy. Ideas that are forced on people against their own will are never accepted by the unconscious. Reason will understand that survival comes first and will temporarily adjust to oppression until the moment comes to have free choice.

It is correct that ideas evolve, but external force is only delaying progress. Only if the unconscious is convinced that an idea has value will it be accepted. But there is a major problem. The unconscious has individual values and a personal identity and Communism in principle represses individualistic thinking. Nature, in time, has asserted itself and individual freedom as positive suggestion of value and from within caused the collapse of the dictatorship of equality of men. And the same fate will happen to the Darwinian concept of the survival of the fittest—an idea that was misconceived about nature and does not belong on the human level. But even in nature, evolution makes sense as positive suggestion of value, in spite of its brutal appearance.

In addition to innate suggestibility, my own awareness of extrasensory phenomena opened my mind to psychokinesis as part of an explanation of the emotional universe. As I have mentioned many times, we are all emotionally connected, but like to remain distinct entities as long as we can. We like to be something unique, instead of being all alike. Eventually it made sense to me, that we are free as long as we play the game designed by nature—to be something of meaning. If we give up, we will suffer the consequences and be replaced be new beings who will play the roles assigned to them. Of course, this means that millions of other humans also want to be something and are in competition for higher positions, resulting in the acceleration of violence.

Historians such as Herodotus, and philosophers such as Heraclitus and Hegel make sense. They said there must be an occasional interference by Providence into human affairs, but not to make life easier or happier for the individual. Rather such interference simply maintains the value-creating process thereby keeping the world a more exciting place in which to experience meaning. The oft-heard lament that nature is cruel leads to unhappiness and disillusionment. Nature is basically suggestible to reason, at least in principle, but the results of suggestion may take some time. Along with several historians, the psychologist Jung believed in the occasional synchronicity of events in history. This is expressed in the timely arrangement of unique persons and events—as if Providence had a hand in arranging our affairs so they would have a beneficial outcome.

From this idea one can deduct that we are free agents but occasionally we are guided. This means the general direction of evolution is set by nature with goals to which humans are useful and expendable. One can assume that the underlying reason is to make the world an interesting experience. When I said that history can be rewritten, it is not the sequence of the events, but *why* they happen, I am looking at what motivates people and the cause of events.

If one keeps in mind that we are all indirectly connected through the collective unconscious, that it is probable that souls never really die, and that suffering is a necessary enforcement to take the world seriously, then it all makes sense. We are in this theater of life as equal participants, with suffering and pleasures to provide meaning. Ancient Hindu and Buddhist philosophy has accepted this idea as truth. Indeed, this could make the suffering in the world a logical part of a larger value-creating process—"indirectly guided" to provide interesting changes.

Let us think of human history as having a purpose, connected to the rise of reason through the expansion of consciousness. Human history started when humans began to think. With the capacity to think, man began to speculate about his role in the world. Some individuals became philosophically inclined and from their speculative reason we have gained partial independence from the fluctuation of nature, have managed to avoid periods of starvation and other calamities that the less-conscious suffer. But a few individuals went further, and with the development of technology and industrialization have interfered in the balance of the nature. With the rise of self-understanding we all must accept the responsibility for our collective actions. Because we have reorganized nature to suit us and did it through force when it could have been allowed to happen naturally, we have disturbed the recuperation and regeneration process. There is also a problem in our perception of time. We live faster, and impact our environment faster than nature seems able to react. This time lag is deceiving us—when nature does not respond instantly, we think nothing will happen from our actions. But things do happen, only in delayed order, usually a generation later or more.

Nature is defensive and guards itself against change. But nature can only absorbs so much misuse until the defense breaks down, like a dam that can only take so much pressure before it collapses. This is typical of the unexpected suddenness by which events can occur. This delayed response is the great danger. We are thinking that all is going well because nothing indicates danger, when nature can break down with catastrophic results. This is also how our body responds; how health is understandable. We may think all is well, and abuse our bodies, when with little warning we become irreversibly sick—it is too late to prevent death.

Biologically, mankind has changed physically but little since prehistoric times, but there have been considerable emotional and social changes. In fact, all major changes involve a sudden emotional crisis after things have reached a breaking point. Therefore, history can be separated by periods that are distinct and reflect the ideas of the people that lived then. This means the history of mankind is reflected in ideas which prevailed in periods—ideas that mostly came from the insight of a few individuals—individuals who inspired and indirectly affected others in a chain reaction or landslide. This happened by innate suggestibility to value, when people saw value in an idea and it became what we refer to as mass suggestion.

The greatest step forward that we made in the evolution of consciousness was to vocally express our intentions and ideas through clear language. With this development, we separated ourselves from inaccurate telepathic communication which is still normal in most of nature. Yet, we have never lost the ability to transfer thoughts—telepathy is only a stage. Our languages can define and describe our ideas better, yet we still understand each other more clearer when we speak face to face, because then telepathy helps to convey thought.

We observe that telepathy is a major factor in the speech development of children. The first language, or mother language, is usually learned by direct suggestion of meaning. Later in life, a level of reasoning is necessary to learn other languages, and to memorize words which is more effort than most people are willing to make. Lack of willingness to learn languages creates ethnic intolerance. Politicians use the difficulty in understanding other languages to agitate people to intolerance and to create hate that will build and grow.

Humans are sometimes called political animals, but this is misleading. We are suggestible to ideas that have value. If we are convinced of the value of ideas, we then identify with them and develop a deep loyalty to them—a loyalty that can be exploited by politicians. Agreed, in nature we all inadvertently exploit each other and, as long as the system is of mutual benefit, it works satisfactorily. A good marriage is founded on loyalty, which includes a certain mutual exploitation. But if exploitation is balanced with reasonable understanding, the marriage can last for life. Since people are usually matched in intelligence, they are capable of equal retaliation meaning that the outcome of conflicts between them is unpredictable. This causes the reasonable person to settle conflicts in a "no-win" compromise.

Some facts that must be clear, before theorizing about human affairs and politics. Value is always a subjective and emotional theme, and one should avoid labelling others. One could be wrong, but without judging values one could not live. That means that one should only judge values for oneself and keep an open mind. Everyone has a personal perspective based on unique experiences. Still we need something to compare ourselves with. This would already answer the range of interest that average people have about their place in the world, namely to compare with other humans. This describes the depth of interest that average people have in acquiring knowledge. Most people only want to live life, not to study life.

I feel it is important, however, to compare oneself with persons of great integrity and insight who have a reasonable outlook on life.

To do so reminds us that values are suggestions which organize into manifest beings and in time, change. When the suggestions that hold an organization or organisms together weaken and fade, disorganization will result. This again confirms that suggestion of value is the reason for all organization because values change in time. Change is a part of the principle of the emotional universe to which we belong.

If many great ideas come from the unconscious in dreams, than the poet Friedrich Hoelderlin is essentially correct when he wrote: "Man is a god when he dreams, but becomes a beggar when he tries to think. When reason enters, the excitement flees. Oh, would I never have known school and science. It has spoiled the happiness to belong to the whole as part of the spirit of the One. Now I have become reasonable and discerning, and differentiate myself from the beautiful dream of my youth and became a single individual, alone in an uncaring world, dried out in the sun of reason."

Hoelderlin is a typical idealist, one of many humans who would like to return to a dream world. But if life were only a dream, without pain, we would not take it seriously. Shakespeare and Goethe had a practical understanding of the meaning of life. Shakespeare saw the world as a stage on which we are the actors in a challenging world and Goethe saw the urge to create something of meaning as the purpose of life. All three poets are right—we are a combination of what they said, dreamers and actors and creators of value.

Man has been thinking in pictures and symbols long before he could speak. Pictures and symbols are the main language of dreams which connect us to the unconscious. This explains why we think and dream in combinations of pictures and words. Written records have only recently been introduced that show in the longest past of mankind, symbols and pictures were used and are still considered important messages from a deeper source—the unconscious.

Given the emotional energy by which things take place in nature, we, the present humans, are mild-mannered descendants of our impulsive forefathers and are on the way to becoming even more rational and civilized. Some of the suggestions that drove the emotions of our ancestors to rage were ideas of justice and truth connected with religion. Those were the common causes to fight each other, because everyone believed himself to be right and felt a loyalty to his ideas.

If philosophy is to be taken seriously it must be supported by comparable parallels in the real world. Prehistoric evidence indicates that mankind descended from smaller animals that were forced to live in trees or hide in caves and had to develop cunning to survive. This cunning was gradually transformed into intelligence and reason. Of major importance to early man was safety—in nature, where might is right, one first must survive. To understand man, the suggestion of safety explains the mentality of prehistoric man.

Humans are called social beings, but not by choice. We are basically individualistic and willingly take some subordination and restriction of freedom for the benefit of the cooperation with other humans that a mutually beneficial society provides. In turn, we have to become loyal to that society. Loyalty starts at the beginning of life. We would not exist if our mothers had not been loyal to us, their children. This is why children are usually loyal to their mothers and family. Early humans lived as nomads and later in tribes in which the males had the main obligation to protect woman and children. As populations increased the defense teams became larger, and organized themselves by electing leaders who established a system of order out of which the feudal system evolved, as military caste. Just as in nature, the early humans respected and admired strength and choose their leaders for cunning—the character of leaders. One better not question their changeable moods.

During the heroic age, security was behind the suggestion of heroic values. If one was strong and belonged to a victorious army, one was fortunate. Yet, if one belonged to a smaller tribe that lost in a battle, that was bad luck. If a male surrendered rather than be slain in battle, he could expect to be executed as prisoner, or used as a gladiator in a blood sport for the victors' entertainment. Fate was kinder to woman and children. They were better fit for slavery and grew up as servants, believing they were inferior. In time, they actually accepted inferiority as their fate. This attitude still partly exists in the unconscious of many humans.

Written history begins with the glorious days of Mesopotamia and Egypt when powerful states emerged and absorbed smaller, weaker ones. In Egypt, the king was elevated to the status of a god, and a theocracy evolved as a dynasty of priests. The king was the divine connection to a spirit world. Besides physical enforcement of obedience by military rule, suggestion of value was natural to mentally ontrol the population.

In Egypt, the submissive population was physically and mentally ruled by a theocracy. It was not yet a feudal system because the king owned everything. People and slaves were assigned to build gigantic pyramids to the glory of their god-king. The suggestion of value which is innate in humans obviously identified with their divine king and gave the people a divine purpose. However, among less subdued people, such as the freedom-loving Greeks, the majority usually elected their government, and people could voice their opinions. Even if they occasionally had a king, as in Macedonia, he was never considered a god. When Alexander the Great built his empire, he was admired as a hero, as selflless, fearless leader, who gave the spoils of war away to make loyal friends. But to traitors he was unforgiving. Then money and gold were used as indirect exchange for goods and services. Material wealth was considered useful, but incidental. To be born a free man, with courage, was valued more because slavery was worse. Alexander was an open-minded king who accepted what had value in the lands he conquered and established relatively fair laws for his subjects. His empire was short-lived and became divided, and the Roman Empire became the dominant power in Southern Europe.

The Romans

When the Roman Empire rose to glory, it was inspired by bravery and heroic ideals. The Romans admired fighters, and the people rewarded victorious armies returned with slaves that they paraded in the streets. In their religion, as in all Indo-European myths, the pagan gods were believed to be scheming among themselves. This was similar to the scheming of the senators and tribunes to become emperors. Although the Romans had a system of laws, their morals were low. In the upper ranks, poisoning and assassination were accepted means to attain power. In our present view, this was an evil empire in a brutal time.

What is interesting is the sudden moral conversion of the ruthless emperor, Constantine the Great, after he eliminated five other contestants in the struggle for the title of caesar. In addition to many other murders, he also murdered his wife and a son, both of whom he suspected of wanting to poison him. What turned this "amoral" soldier of fortune around to further the cause of morality? Was it because of a dream? This is what he gave as cause for his conversion to Christianity. In this dream, Constantine saw himself winning the world with the Christian cross on the shields of his legionnaires. But I think his conversion to Christianity was more likely based on insecurity—the realization that one cannot trust anyone, not even his own family. He wanted that trust and loyalty should be instilled in mankind.

This insight was the beginning of a change in European history. Constantine understood that brutality could not go on forever—there must be a gentler way to live—and he became convinced that a sect, preaching love, had the right message. He halted the persecution of Christians and made Christianity the official state religion. He built churches and furthered the establishment of a church hierarchy as a parallel to the worldly hierarchy. From that point on, the hero image in the Roman Empire was replaced by the image of the suffering of Christ on the cross. Inadvertently, the empire declined.

After Constantine, the empire divided and there followed five centuries of turmoil during which the western Roman Empire and papacy nearly collapsed. Then, a Teutonic chieftain, the king of the Franks, Charlemagne intervened as savior and restored the authority of the church again. He accepted the title of caesar and accepted the role of protector of the Roman Catholic Church. This begun the tradition of the Holy Roman Empire with Germanic kings calling themselves kaiser, which is a phonetic version of caesar. Similar to Constantine, Charlemagne also wanted to turn the savage Teutons tribes into a gentler, tolerant people. This started a historical period, wrongly called the Dark Ages. In order to do achieve the conversion to Christianity, Charlemagne had to use force. The pagan tribes had heroic ideas and a long established religious mythology. Many resisted and were eventually converted after witnessing mass executions. Charlemagne, whose name means Karl the Great, built churches and schools and had Latin introduced as the common language of the empire. Charlemagne also adopted Roman laws and installed courts of worldly justice. With the repression of paganism, Charlemagne, with the help of the Church, started the civilization of Europe north of the Alps. This was the period in which black-robed monks, with crosses in-hand, instilled love and fear for God and turned Europe into a civilized place were one could live in relative safety. Absolute safety does not exist, not even today, but the idea of a gentler way of living found many followers, and loyal defenders of the ideals expressed by Christianity.

Humans by nature are peace loving, and if possible would rather trade than rob and steal. With law and order established, religion began converting the people to become pious citizens. Trade and commerce flourished and a monetary system of exchange made all this workable. The military caste took advantage of this improvement by charging taxes. In fact, the military caste, that gradually became the landowning aristocracy, had made the people into obedient subjects and installed sharecroppers on their large holdings to work the land for a small part of the yield. In general, the aristocracy became a feudal caste, living lavishly from exploiting their subjects as much as they could until the people revolted. Originally the upper class came from the people, but over a few generations had distanced themselves and intermarried with other aristocratic families to increase and preserve their landholdings and tax base. Thus the aristocracy became an inter-European class to which ethnic difference and nationality was irrelevant.

This period of history is called the Feudal Age. States mostly appeared through the acquisition of territory by marriage (love had little to do with it) and by politics. Occasionally land also was annexed by invasion. An example was the Hapsburg family which was originally German but had intermarried with Spanish and French families, and many other nationalities as well. The aristocracy became estranged from their own people. To speak of the Feudal Age, one has to think of an international class to which the interest of the population was irrelevant. This reminds me of international business interests in our time which have little concern for the welfare of their nations or workers, and loyalty is onesided—expected from lower class. Let's go back to the Feudal Age and the Holy Roman Empire during its end stage. Besides the feudal lords, there were the lords of the church exploiting the population, using the military to force obedience. When revolts did occur, the nobility with its trained armies could easily defied the peasants.

The intellectual dominance of the church was challenged when, during the Renaissance, other ideas entered Europe via the defeat of the Moors in Spain. The translated books of Greek philosophy contradicted the truth of the *Holy Bible*. The church had lost its idealism in the highest ranks and calls for reform came from Martin Luther and others. Kaiser Rudolf II was tolerant and ignored demands from a new pope to suppress the Protestants with a military campaign. But Rudolf was replaced by a younger brother who gave in to the pope's demand, and the "war of the counter reformation" began. After thirty years, general anarchy made the continuation of the war a futile effort and peace was finally made. Religion was declared a private affair and the church was separated from the state. With this declaration, the Holy Roman Empire officially ended.

Religion became a private choice and radical sects who would not accept the liberal decision, emigrated, mostly to America, were they could follow their strict faith. The feudal system remained in control, but the general trend was already towards liberalization. Commerce and industry developed and cities elected their own administrations and demanded concessions from the upper class. But the differences grew deeper because the aristocracy still had a privileged status—above the laws that only applied to commoners. The spread of literacy made possible by the use of the printing press influenced the population to demand equal rights for all. This situation existed until the French Revolution broke out. It was in France, where the feudal system had reached its limits, that life for the lower classes became unbearable and the arrogance of the aristocracy angered the people. A revolt finally was started by a few idealistic intellectuals in Paris who believed that justice should be equal for all. It gained critical momentum, even spreading to the soldiers who guarded the king. When the guards broke their loyalty and sided with the people, the revolution succeeded. Eventually the king and queen, along with many other aristocrats, were beheaded. This was the beginning of liberalization and the gradual decline of the feudal system in Europe

The beheading of so many nobles in France scared the aristocracy across Europe and they raised an army to restore the French monarchy. But this army was defeated by the elan of the people's army, which beat them back. Then the French, encouraged by their success, intended to liberate other lands which they actually conquered. Many in Europe now thought they would also become liberated. However, Napoleon, a general of the French army who led victorious campaigns, had become a hero to the people and used his prestige to crown himself emperor of France. This proves that people admire and glorify leaders. This started the reversal to aristocracy and feudalism. Yet, Napoleon had enough sense to introduce needed reforms in government and law to appease the majority. For the rest of Europe, the revolution was a disappointment.

The idealism of the French Revolution was not all in vain; other monarchies introduced similar reforms to avoid a revolution. Parliaments were established so that the wishes of people would be heard. The upper classes still defended the feudal system and the working people turned to socialism and formed the opposition. With this event, "national concerns" became paramount and the aristocracy had to accept that national interest were overriding their private relationships. A new period of nationalism entered history. Religion still had a lot of influence on the minds of average people. Now the differentiation was according to language, which meant that ethnic intolerances replaced religious intolerance. In fact, religious and ethnic differences still cause violent clashes in our time.

The Industrial Age

The details of the changes that occurred are interesting for historians, for a general overview it is the transformation of the perception in the people telling a story of the expansion of consciousness. The Industrial Age turned into commercial competition for markets in which nationalism was used to fire the emotions of populations to justify war in the national interest. Religion became of secondary concern, usually surviving in rural regions. The nation states of Europe experienced social unrest, inspired the rise of socialist parities that demanded a share of the profits of their employers. Materialism and discontent belong to the consciousness of the twentieth century. Individual materialism is the new religion of the majority of people. Socialism could not survive against the natural need to remain individuals. In time, all efforts to regiment people dissolved because people wished to control their own lives. The development of reason in the greater population is an indication that we are moving to the acceptance of individual differences and tolerance. The new issues will focus on finding a meaning for life of the ever-expanding population because mechanization is counterproductive. What will all the people do when automation and robotic progress further eliminates workers? We are certainly moving into a new state of human history where the meaning of life is an issue, not just hunger and survival. At least with democratic institutions, peaceful transformations are provided, and with foresight, catastrophes can be prevented.

Eventually the world is raised to another level of experience, on a different mode of being in which the physical infrastructure is secondary and other values are current. We will not find out what the new era will be because such changes take time.

This is a deduction based on the structure of levels that exist everywhere, but still only a deduction. It is possible for us to live together on this planet if we can communicate and accept decisions based on fair judgement. This would be preferable to violent means which usually solve nothing. The war of reformation solved nothing. It is up to each individual to decide what to believe. The French Revolution solved nothing. Napoleon did reverse democracy temporarily but democracy was inevitable. It took years for people to become used to freedom from oppression, and to believe in equal justice, which is difficult to establish and satisfy in the best of worlds. That is what nature has intended—we all want and need to be something special and to suggest value.

Our current problem is obsession with material wealth which is futile because we can only have temporary ownership of material goods. Ultimately, we own nothing but our consciousness. But let us think about materialism which is with us as our current problem. The idea of equal distribution of material goods is old and yet, forever new.

Let's go back. When after the first great war, in which Russia was defeated, and Vladimir Ilyich Ulyanov, known as Lenin, a Communist idealist, used the opportunity to proclaimed a socialist dictatorship and with this act began the greatest social experiment in human history. It was an experiment that spilled over to China and other places around the world. The dictatorship was meant to force "equality" on more than two-hundred million people who by nature were all different individuals. Everything worth owning was declared communal property, causing violent opposition and mass executions. This reminded me of Karl the Great when he civilized the Teutonic tribes, but he never intended to make them all equal or to disown their property.

The communistic system is based on economics, on the equal distribution of goods but this also demands "personal equality" among people, degrading the purpose of life to produce and consume in a strictly material universe. Life ends at physical death without the hope of survival of consciousness. Communism rejects all spiritual phenomena and is totally based on scientific materialism. Yet, the communist ideal, in the course of seventy years, turned to jealousy because humans are different as individuals, and denounced each other as traitors. Brothers and sisters they were not but frustrated humans who longed for freedom from oppression that had been established by an unrealistic idea. Pride of achievement and admiration that we all need was considered elitism. In short, the system repressed the individual experience of self-worth in order to assert equality. It was inevitable that Communism would eventually collapse. Obviously, nature has already invented the best system when it created individuals with minds to think for themselves instead of having a central government to tell them what to do.

Let me mention some personal experiences. When I came to the age where reason begins, at about fifteen, I left school and started an apprenticeship in a factory that built machinery for coal mining. The year was 1939 and the Second World War (which was supposed to correct the injustices of the First) had just begun.

I had been a daydreamer in school, but now, faced with the reality of life, I saw the need to either educate myself or spend the rest of my days in noisy machine shops, working for a living the hard way. I was lucky to get a chance to learn such a trade, which gave me the possibility of studying mechanical engineering later.

Besides the technical training, in evening classes I took English and French language courses. In the factory, I was moved through different departments. In one department I was assigned to a Czech journeyman. I remember him well because of our long talks. While eating our sandwiches during lunch breaks we came to talk about philosophy. He knew about Socrates, who said that we know only a small part of reality, and that most is unknown. We agreed. He also was a pacifist and deplored that another war had occurred because the common folks were losing their lives while the rich were making profits. His ideas were along Marxist lines—if the working people would unite and control the distribution of all goods according to need, then all the conflicts over material possessions would end, and the need for war would disappear. All humans should have equal rights, he thought, because we all have common needs. This meant that Communism was the only hope for a peaceful future of mankind. My point was that we humans may have similar physical needs, but individually we are all different and do not want to be the same. If all are the same, there is nothing interesting left to live for. We agreed to disagree and respected each others views because in principle we both were socialists.

In the fall of 1943, when the war had already taken a reversal, at the age of eighteen, I was drafted into military service. The German Army was retreating in Russia and the Americans had landed in Africa. When I got to the Russian front in the winter of 1943, it seemed that the tide of the war was already irreversible. I contracted an infection in one foot from walking through mud, and needed an operation, after which I was transferred to a hospital in Austria.

After my recovery, I was given a furlough for two weeks at home in Bohemia, but I had only been home a week when I received a letter ordering me back to a garrison near Berlin. Nobody seemed to known why, but I soon found out. There had been a plot to overthrow Hitler. A bomb had been detonated in the command bunker which served as strategic headquarters for the generals. Hitler had survived the explosion, but in preparation of the plot, and in anticipation of success, the revolutionaries had sent out orders for available troops to march on Berlin. The insurgents had also sent a regiment to take control of Radio Berlin to announce the death of the Fuehrer and the formation of a new provisional government to end the war. The officer who took over the radio station was already waiting for the arrival of a speaker who failed to arrive, but instead received a personal telephone call from Hitler, saying that he was still alive. By appealing to his oath of loyalty as soldier, Hitler persuaded the officer to follow his orders, which was to announce that the plot had failed, that the Fuehrer was alive and that all troops were to return to their garrisons. I can only imagine that this officer did have a conflict between reason and loyalty. His reason was telling him that prolonging the war was a useless waste of lives but he would also committed treason by disobeying an order from the highest in command. His loyalty won over reason and Hitler was back in control. The consequence of the failed plot was that the war continued for another year. Many more people were killed, and many German cities were flattened by bombs. Finally, when the Russians already were in Berlin, Hitler shot himself.

It seemed that Hitler was indestructible, as if providence had plans, but not to help the German people. It was as if the mission of Hitler was to destroy Germany when he thought that unifying Europe would be his mission. Perhaps total destruction was necessary for a new Europe to rise from the ashes—nationalism and all traces of militarism had to be obliterated completely. With this logic, it means that Providence does not count lives. Millions of innocent people had to die in horrible circumstances—people who had little to gain from the war, one way or another.

Let me return to that point when Hitler survived the assassination plot. He returned to Berlin and ordered a thorough search for the persons responsible. Using torture, forced confessions revealed the names of other conspirators before those involved in the plot were killed. It turned out that many prominent persons were involved. The list included some aristocrats, and the son of Planck, several mayors of big cities, also many officers, even Field Marshall Erwin Rommel, commander of the western section. And Rommel was the second reason why it seems that Providence was involved. Rommel was on his way to Berlin when an American fighter plane fired on the car in which he was riding, seriously wounding him. This was the reason why he did not arrive at the radio station in Berlin to announce the end of the war. He was chosen because he was a well-known senior general and had the personality to control the military and speak to the people. Not only did Hitler survive the attack, but Rommel was prevented from ending the war. Was all this just coincidence or was this interference by Providence, in human affairs? This idea was expressed by the Greek historian Herodotus, by the philosopher Hegel and the psychiatrist, Jung. Jung in particular noted a phenomenon that he called synchronisity of events, prearranged to coordinate the outcome of major events. Synchronisity occurs not just in human history but in the life of ordinary individuals, when people meet in circumstances that seem to be intentionally arranged rather than happenstance.

In hindsight, now I can see why Providence was against the conspirators, all of which were prominent personalities in Germany, to restore Germany as a nation. They were eliminated because a new era of a United Europe was being prepared, in which the revival of nationalism was an obstacle—perhaps with a united mankind in the future in mind.

Such thoughts I have now as a seventy-one year old man who is looking for a pattern in history. But when I was nineteen and was hoping that the war would come to an end, I was disappointed that the plot had so miserably failed and the goodwilled men were executed, while at the same time the insane leader was protected by some uncanny, inhuman power. Religious persons would say it was God's will and that we should not question God's judgement. But what about the millions of humans that suffered injury or lost their lives? I am certain that many shared this opinion because deserters were frequently found and unceremoniously hung by the neck. They were left hanging for others to see what happens to deserters. It was the trust of average people, loyal to a common cause that extended the war for another year. In that time, the bombing of Germany intensified, resulting in the interruption of the transportation system and causing starvation in the cities. Hungry people sold whatever they had for food, mostly to farmers or profiteers. But those who could not get to farms suffered the greatest starvation—persons conscripted to work in factories or locked up in concentration camps. One can say that the bombing not only hurt Germans, but indirectly large numbers of foreigners working in the factories and local women starved. Because all able-bodied German men were already drafted into the war and many had lost their lives, boys of sixteen were enlisted and even old men of seventy. During the war's final year, there was a waste of lives which defies reason. When the war finally ended, thousands of prisoners of were liberated, and concentration camps opened. Many prisoners had starved to death in the last days of anarchy. This inhumanity was probably unintended and was indirectly caused by the destruction of the infrastructure by indiscriminate bombing when the war was practically lost. I am not denying moral responsibility, but as an individual, one is powerless.

What lesson can one deduct from this tragic event about human nature? It was a confrontation of loyalty against reason, an inner conflict in the soul of many Germans. Yet, the higher powers that guide history were supporting the conservative side of loyalty, regardless of the cost in lives and how insane it all was.

Let us calmly think this deduction through. Loyalty is the most important value in life, perhaps more important than reason. What is life without loyalty? One has to stand for something. One must be loyal to oneself to remain sane. One must have values or one is nothing. One must remain loyal to the wife that one has chosen, and to the children, to give them a chance. One has to be loyal to people with which one works or the order of civilization would collapse. Without trust, human society is inconceivable. One has to trust that our food is not poisoned, that our monetary currency has value and so on. Trust and loyalty are basic in human affairs, but can never be unconditional. Reason must also prevail. Without reason, we would all be insane. Because humans are flexible to changing conditions and because we can reason, we humans have the upper hand in nature. We expect loyalty from others. We trust the planets are loyal and our sun will shine a long time. The sun keeps shining and the planets follow their path. Because we expect to live tomorrow, that tomorrow always comes.

If all things are following one common principle, it makes sense to think that it is loyalty that keeps the planets in orbit and holds the world together. On the highest evolved level of human experience it is loyalty that is important. Even on the physical level, the attraction that physics calls gravity, in higher conscious experience has the same effect as loyalty has on the idea of creating meaningful worlds, which gives purpose to the value-creating process.

In fact, in nature each species is loyal to its own kind. And yet, there is also reason and progress in the evolution of nature. But progress happens slowly and is best achieved through mutual agreement, even if it takes longer to evolve naturally.

It is apparent that loyalty is a basic fact in the stability of nature, but so is reason. Reason is necessary in the process of evolution by meaning. Reason has evolved to an equally important degree on a conscious level in human beings, but is basic on the unconscious level, particularly, in the educational system. There is sometimes a conflict between reason and loyalty. This occurs when loyalty becomes dogma, just as loyalty to ethnic values can hinder peaceful cooperation among people.

There were many reasonable people in Europe who saw the meaninglessness of divisions by national interest and have established the idea of a "European Union." This began as common market for peaceful exchange of goods with a later integration as one political entity. Similar trends for free trade worldwide could eventually lead to a world federation during the time when technology has already turned the world into a global village.

The end of the heroic era happened in Europe, and was forced on Japan and the rest of mankind by the development and spread of atom bombs among the major powers. Because of the awesome destruction of nuclear bombs, war became suicidal. During a period of posturing, called the Cold War, it took a long time for this concept to sink in. We now realize that nuclear war is senseless and that the only possible future for mankind is to live together in peace or not at all. Human history is forced by reason into peaceful coexistence, but industrial mass production has caused concern. We need to rethink the direction of industrialization as an ever-accelerating growth factor.

We must ask: Can this planet absorb unlimited expansion? There is a limit to the amount of pollution that nature can absorb before it suddenly breaks down, which is how nature works. If hydrogen bombs did not destroy the biosphere, pollution will—when we reach the top unable to know where the top is. Because of the delayed action, a point of accumulation will be reached when nature breaks down suddenly. Therefore, people do not recognize the danger until the ecological catastrophe is irreversible.

We seem to speed towards a self-created ecological catastrophe. It is as if we are forced by reason to change direction, to become responsible, or cease to exist. While developing better mechanisms, we overlooked that the universe is an emotional and conscious process? To understand nature, one can see its intent best in the highest evolved stage, in human affairs. This means suggestibility and organization extends through all stable materials and the ground on which we walk.

The reckless competition for the same material goods is misleading most humans. We do not need all the material things. Happiness is a suggestion of value within our self. Money cannot make one content. Any role in the world is something important. To experience meaning is the issue. Materialism is leading mankind to increased discontent, driving many into acts of violence and crime simply to posses things that belong to the material infrastructure—

things that one has to leave behind when the soul departs. It is true that the body cannot live without food, but food is not enough. In order to live one must see value in life. Psychology reached this conclusion a long time ago, and it is time that we connect this knowledge with the universal psyche, which is the primal essence of nature.

Concerning our species, we are undergoing further separation and reorganization which could be called "individuation." We have improved human life expectancy by technology, and thereby have interfered in the organization and balance of nature—inadvertently nature became our affair and responsibility. When technology started to crack the atoms of matter, it became necessary to understand the meaning of reality before we do more damage in a one-sided drive for mechanization. We should revise our thinking. Nature has given us many signs that mechanization and reckless exploitation are the wrong direction for intelligent humans to pursue. Indeed, our planet has shrunk and has become a "global village." It is a planet that is becoming overpopulated and we are responsible for our action because we have evolved to reason to know what is happening. Everyone knows that population control is necessary, but in an agreeable way. If one is loyal to old beliefs that more children are better, then one should also live in that mode to which that belief applies. Starvation, sickness and war controlled the birth rate. But in our modern world, we live a sheltered life, and we believe that one should have fewer children because almost all will survive under present conditions.

As imperfect as life may be, it is still more interesting than the darkness would be without the light created from emotional excitement that makes the world an interesting place in which to experience meaning.

9

About Individuation

HUMAN CONSCIOUSNESS IS DISTINCTLY INDIVIDUAL. In fact, all awareness and perception is sorted by the subjective interpretation of individual experience. Most of our civilization, its art and literature, is packaged for individuals. Even if knowledge is presented in abstract generalizations, a form that is invented and artificial, we are beginning to organize our world view with anecdotal thinking from our experiences. Even academically trained persons, who in their professional work use abstract thought and develop mechanical formulas, in their private lives will return to emotional living in existential reality which is individually experienced.

The general pattern of physical evolution is unconsciously organized by innate suggestibility to a multiplicity of ideas, which are the meaning by which different species exist, and are expressed best on our organic level. On the organic level most beings have only a narrow range of awareness. It is on the human level that this narrow consciousness becomes expanded through experience in a human body. A complex organic body has been organized for human consciousness to become used to available functions that the body can provide. The five external senses, and the voice box that allows speech to be learned, and the brain that allows experiences to be organized by meaning, are available for the consciousness to use. These are advantages that lesser evolved species do not have.

In short, we are born as a field of conscious emotion into a physical structure that provides many functions and limitations. During early childhood we learn about these functions and our limitations.

Consciousness expands as it adjusts to physical existence in the body. Then, consciousness adjusts to the existing social structure in a human world. In principle, this is a process of expansion of consciousness provided by a structure which, in slow physical evolution over countless generations, has evolved as organic body. This means there is evolution within evolution as the expanding individual self-consciousness occurs in a subprocess that one could call "individuation" of consciousness. During a person's lifetime, one can acquire a unique identity. By "can" I mean that this can occur if one happens to be born in a supportive environment. A supportive environment will help develop an individual identity. Academic learning or twentieth century technology add detail, but stimulating life experiences are essential in expanding consciousness.

Under normal situations, if life is interesting and the individual is content with himself, there will be a functional cooperation between mind and organism. Happy persons are usually healthy and their various organs will perform their functions unless physical or emotional conditions challenge or disturb the process of individuation. This process is mainly a form of maturation from experience in life. From an experiential perspective, the mechanization of life through technology is merely a convenient enrichment, a sideline that leads nowhere. In fact, we have to reduce our demand for consumer goods if we want to maintain the organic level on which the body depends without causing irreversible damage to the balance of nature. When I said in the first chapter that we are going in the wrong direction, I meant that science and technology are being misled by a materialistic concept of reality that is mechanical. The universe is an emotional experience that evolves towards greater quality. The expansion of consciousness is a subprocess in a mental mode. This is the direction we humans are following. Philosophy has followed this direction until mechanical methods were introduced. Ever since, a large number of educated persons have been drawn to mechanization because it allows the production and exchange of goods, an artificial mode of mechanical functions. Yet, the physical side of reality is only an infrastructure to the experience of value and meaning.

In this century, psychology has begun to rediscover consciousness, but from negative consequences of the emotionally disturbed functions between body and consciousness. Psychology found specific causes as to why discontented individuals give up on life and punish themselves and psychologists have found that they sometimes could talk suicidal people out of giving up. Although the psychology of the unconscious did not directly name innate suggestibility as cause, I have found this to be the common cause which applies to all problems. Because the cells of the body are suggestible to value, for whatever reason, the unconscious can restore cooperation again, if it can be convinced that life has value. That this is true is experimentally verifiable because psychology has healed people by suggesting value. This happens "indirectly" when the individual reaches the insight. It happens as self-suggestion (catharsis).

That suggestion can also influence the body beyond the restoration of normal health is confirmed by yogis who can influence the cell state to comply to unusual orders, regardless of religion. The yogis' influence goes beyond hypnotic suggestion. Should we now want to become yogis to influence our body? This question has a practical answer, because total self-control over the emotions demand ascetic living which contradicts the purpose of enjoying life and the experience of meaning. Since yogis want to suppress all desires, because desires cause suffering and pain, they also cannot enjoy the physical pleasures of life. In meditation yogis and Zen masters try to connect to the source of the undifferentiated consciousness, which is presumed to be existing in nothingness. If achieved, this state offers the liberation from useless suffering confined to a body in the cruel world. Yet, the effort of evolution goes in the opposite direction—to create interesting worlds of meaning by expanding consciousness. This happens by differentiation and the creation of polarity of values. The conflict of opposites does cause pain but also pleasure. One can reduce the suffering of pain with intelligence and with foresight, can find that life is worth living. Mankind is evolving selectively in this direction on an individual basis.

The differentiation and polarization of values only sometimes causes pain and suffering, when specific ideas of value are opposed. But life is worth the price even when we must learn to live with opposition. The experience of meaning is more than nothing. It is as if nature wants to get out of boredom that existence in the peaceful void implies. This peace may be what disillusioned individuals seek at the moment of hopelessness, yet may find out it is uninspiring after a while and return to a new experience of meaning in a physical form.

Suffering is also an emotional suggestion. It is mainly a warning signal. Pain need not be over emphasized because we can avoid pain by intelligent living. The efforts of man to reduce suffering are working. We are combating hunger and disease through medicine. But some reminders of pain are needed still, because life has to be taken seriously. Even the suggestion of pain, or the constant reminder of painful death that can occur when someone is careless or takes risks is a warning that pain will assert itself and life can be lost.

Having assumed that there is a common principle for everything as an explanation for evolution, there is then only "one" acceptable motive to explain the material and spiritual aspect of life—namely innate suggestibility to value. This principle is capable of generating emotional energy and ideas that organize complex individuals to reach self-consciousness—conceivable within actual experience. On less evolved levels this intent may not seem as obvious, which is why materialism appeared. But on the human level of evolution, it becomes irrefutable. In fact, the direction of evolution is toward meaning and toward living in harmony with nature and the value-creating process. This is a better scenario than resenting life through a negative interpretation of reality.

One can observe in the patterns of nature that evolution has a direction towards individuation. Nature has tried to create diversity through endless experiments. For example, among insects matriarchal societies are common. The hives represent one single family all descended from the same mother. All identify with the family and are loyal to the efforts of their mother. To keep the family functioning, there is no place for individuality in such an organization.

In a higher mode of organization, as for instance among mammals, greater individuality exists. Mothers still give birth and raise their offspring, but they must teach them to exist on their own, to break away from them to live as self-sufficient individuals. We humans belong to this pattern. Normally we want to become something by our own effort as individuals. This intent results in greater creativity and separation—each new generation of individuals assumes a separate identity from their own experienced values, which are different from their parents. In the process of differentiation, reorganization is involved and the clear intent towards individuation is also expressed. This confirms the value-creating process towards self-consciousness and independence as individual.

Nature has many levels and modes of being, yet even on the atomic level and among the stars, which are massive congregations of suggestible entities, simple organization must be assumed. Stars have separated from galaxies, and planets have separated from stars. In both cases they travel around their mother, but keep a distance as separate but loyal children. Physicists would claim that gravity holds the planets in their orbits, but I dispute this. Mechanical gravity would not allow elliptical orbits and the gravity, as we think is necessary to hold us and all things down to Earth, is only a narrow field of attraction. Instead, planets and stars are influenced by suggestiblity to value. Outside this so called gravity field, man-made objects move in a straight line, but stellar members move in orbits. This is explained by the theory that loyalty to their mother and allegiance to family are values that causes the planets to keep regular orbits, regardless how far out into space they may swing like Uranus and Pluto. On the organic mode, such family loyalty is found in many insect species. They may fly far and wide but return to their own hive and no other.

The idea that suggestion of value as loyalty holds the organizations together is universally provable. It recurs with organic and inorganic organization alike. Therefore, we will have to take notice of this common fact. My initial intention was never to prove anyone wrong, but if there is only one single cause, all other causes are only secondary to suggestion of value—as everything else in the universe appears to be without exception—then many must be wrong.

Certainly, the innate suggestibility to value is also confirmed by human psychology that deals with the unconscious. The psychology of the unconscious leads into a strange field of imagery, which relates to suggestibility. This can actually heal people of certain emotionally based illnesses that respond to suggestion of value. Other illnesses, that do not respond, are on other modes connected to their particular identity.

Consciousness on the organic level is a basic state of emotional alertness and expectation. This awareness is directed forward to expand to greater self-consciousness in the expectation of an interesting future. This keeps us mentally and physically in expectation of a future. It is the life force or the *elan vital*, that is presumed by many philosophers to be the essence, behind which I think, suggestions of value organizes and sustains life and remains viable as long as the organization has a purpose. On the human level, psychology is appropriately called "individual psychology." The further individuation of consciousness continues when one becomes independent and judges value for oneself—when one thinks for oneself and becomes a reasoning individual. At this point it should be recalled that reason is the same as judgement of value and meaning.

In details, each individual is unique, but in principle, all humans have many common values and needs. Therefore, normal individuals mostly agree with the majority of general human values, because individuals posses "reason" to understand our place in the scheme of things. Some persons may want a further definition of the word reason. It means a personally acquired judgement of value from lived experience. This is a conscious, flexible interpretation of reason—nothing like what mathematical logic expects reason to be—absolute and constant. In life, one is never completely certain. One must consider values and priorities in constantly changing situations. This is where humans are superior over computers and systems of logic—we are flexible and reasonable beings.

The idea of a conscious world-soul was mentioned in Sanskrit writings of more than four thousand years ago. The world-soul is called "atman," of which kundalini is the living spirit in all of us, supporting the claim "you are that." Yet who knows how old this idea was before it was written down in the *Rig Veda* of ancient India. Many ancient philosophers never wrote anything. Consider the situation of Socrates. Plato gets credit for many ideas that were really Socrates' because he wrote them down. Regardless of who had them, ideas have a strange way of coming into consciousness. Some speculators think that knowledge evolved as a stream of human thought—that is accessible to anyone serious enough to make an effort to know. Then ideas come as suggestions, in symbolic form, as general advice.

The history of philosophy is a record of many transcendental insights gained from the "unconscious psyche of nature." Individuals had ideas which at their time were unsubstantiated but in our present time have been confirmed. When psychology is the subject, everything becomes individual, because it deals with the intimate feelings and private experiences of persons. Although philosophy was the precursor to the discovery of the unconscious, psychology added empirical knowledge to the understanding of the human psyche, which at least in principle is a replica of the larger psyche of nature.

The professional experiences of pioneers, such as Sigmund Freud and Carl Jung, with mentally disturbed individuals has defined many details of how the unconscious perceives values, in general meanings. We must learn from psychology and the personal level of experience of real examples from life. And if a writer cannot exclude himself, then that must be taken into account.

Let me mention an example. I have to earn a living like most people do, and I also talk to other persons if they want to know my opinion. One blunt man said to me, "If you were really smart you would not work for a living." I still owe this man an answer because I did not know what to say at that moment. It is not that simple because we cannot all have high-paying positions or have our own businesses. Most of us have to work even though we may be overqualified for our jobs. There are not enough positions open, so chance is involved. In the future the job situation will be worse because as more people become educated, most will have to accept manual work to earn a living. Besides, to take wealth as measure for intelligence is simplistic. Many came to their wealth by luck or cunning, if not by inheritance. More important than gathering money is to have an interesting life and to be content with oneself. Of course, there are many simplistic people in the world who think wealth is an indication of intelligence. They will be disappointed. If they personally knew some of the rich and famous people they might change their opinion. Most of the wealthy are just narrow-minded persons with money invested to make more money.

In general terms, we can begin with the fact that all humans are protected from outside interference by a closed identity. This makes us only indirectly influenceable. To be influenced in deeper functions one first must be convinced of the value of an idea. But before it can take root in the unconscious, we indirectly suggest it to ourselves. To self-suggest it as of value, the unconscious believes only in itself and personal experience. This fact is confirmed by psychology and psychiatry. If a patient understands the reason for a persistent physical or mental problem by his or her own insight, the problem will be resolved and the person will become well and normal again.

Individuation means to become mature, meaning that one grows in self-understanding. Nobody else can bring this about; one must be mentally ready. There is no guarantee that growing older, or becoming physically mature has real meaning and the unconscious has gained in self-understanding, except through results. This is also why the dialectical principle misses the point. Force only convinces persons to avoid suffering, it does not cure. We cannot be assured of the ultimate success in the reform of criminals. We can only know if that reform worked when a person actually lives a different life without relapse into criminal behavior. Verbal assurances mean nothing. Criminals may not mean what they say and may risk going back to prison in order to get rich fast. In other fields, such as personal relations, some persons stay immature their entire life. This is evidenced by recurring failure in marriages, and is proven statistically by repeated divorces.

But as civilized humans, we also cannot follow our inner impulses—at least not directly. We must control and defer our natural drives to a proper time and place. So we learn to control our sex drives, play down our self-worth and avoid bragging. Because everyone can have opinions, unless they are confirmed, opinions remain only talk. Over time most of us eventually change our values, because as children we had simple ideas and as adults we learn to live by reason.

We are born either male or female, about which we had no choice nor did our parents. We also are unequally endowed in physical appearance and stature. In addition, the astrological position of the planets at the moment of birth affects our emotional self and seems to have a subliminal influence on our mentality and talents. All these factors mean inequality from the point we enter into life and have consequences on our chances in life. This means life is intended to be "unequal" from the beginning, and during life, this inequality continues to increase. We become estranged from earlier friends and then associate with people in similar situations closer to our experiences and values. This is a normal reorganization of identity which makes us change values to acquire an individual identity.

Inequality means the world is meant to be challenging—physically and mentally—to facilitate change. Other factors such as wealth and education are less important—actually a burden to develop a unique identity. Wealth creates false personal values and eduction wants to instill conformity in thinking. One is told what books to read and what methods to use to arrive at expected conclusions. But later in life intelligence sorts itself out and the motivation to acquire knowledge for oneself reorganizes the playing field. But the actual differences are unpredictable—status and class is only for unthinking persons. If one could know in advance, there would be no surprises. I conclude that this implies that life is meant to be interesting because all is uncertain.

It is clear that nature never intended equality to exist, but rather to provide for the dynamics to create interesting experiences. If one wants to believe in justice, and fight for equal rights, this may give one a purpose. But it is a hopeless one to fill. One will eventually come to the conclusion that imperfection is the reason why the world has always something left to be changed and improved on—eternally.

If it were otherwise, and everything was goodness and loving, what would we do with ourselves? With nothing to fight for and nothing to set right, life would be boring. This world, with all its problems, is the best of all possible worlds, just as Leibnitz said it was. Voltaire was naive, but Voltaire had fun fighting for the rights of man. Life gives us something to set right, a job that is never done. This sounds like the "continuum infinitum" that the universe appears to be from a general perspective.

One can be immature and cry about the injustice of the world, or fight like a Don Quixote against evil. Or one could reach maturity and think for oneself. Individuation begins when one understands what is real and cannot be changed. If one comes to such an understanding, one cannot be disappointed. In fact, there is a choice. If one dislikes confrontation, one can avoid it. In this way one is saved from becoming involved.

In nature there are many entities which have taken safe and uncommitted positions. They can be classified as introverts or referred to as autistic and incommunicative—a common characteristic in plants and material objects. They live within themselves, withdrawn and unresponsive. Unfortunately, disillusionment occurs in depressed humans. Because as a human, one is committed to provide for a demanding organism that wants to survive and experience interesting adventures.

The polemics in the world arise from the polarity of values. It means that human existence is mainly extroverted and is meant for action and participation in the world. Withdrawal on this high level of extroverted consciousness is not a happy option. It results in a throwback to lower levels where introversion and autism is normal. On the human level, a balanced mentality is considered normal and extremes are viewed as abnormal. Speaking of mental illness, the opposite of depression is mania, which also frequently occurs in frustrated persons. In some cases, mania and depression alternate in cycles. In a general way, these emotional responses are pointing to unconscious conflicts by expressing dissatisfaction with the conscious level, thereby causing mental confusion that the external world views as insane.

The normal, healthy individual with a positive outlook has also been studied by psychology. The self-actualizing individual is that person—evolved on a secure level of self-assured worth with the body and mind cooperating. Many problems that children have are rooted in the insecurities of their mothers during pregnancy. These insecurities were transferred emotionally to the fetus and infant. On the positive side, secure mothers in a stable family setting suggest value to their children because they feel valued. Unfortunately, many family situations involve conflict or divorce so that this experience of being alone causes increased insecurity. This leads indirectly to more criminality and mental illness in the next generation. The majority will be followers and conformists, who leave thinking to others because they are unsure of themselves. Such insecure persons have a low level of trust, and see the reason for their problems in the malice of others, rather then the unfortunate circumstances in their early life.

No doubt, many individuals have reason to feel insecure and discontented Since we are suggestible beings, our imagination can run wild if we lose control over our judgement—if we let it happen. By just looking at other species in nature one can understand that they have restricted their imagination and avoid comparing their value with others. This can only lead to feelings of discontent about the injustice in all of nature. In this sense, we humans are favored. Compared to other species, to be human is something valuable. If one were to compare oneself to other humans there are differences in value. But this happens only if one compares oneself to others, thinking they are better off, richer and more advanced. This causes feelings of inferiority. If one is content with one's own value, one knows that others are only differently organized and have their particular role to fill. Each human has to carry his or her own burden in life. What may seem desirable to others is just a burden one may not want. The rich have the burden of their riches to worry about. Occasionally, some drive themselves insane about losing a few million and think this is the end of the world.

Most beings in nature are only concerned about how to get by from day to day. However, the great majority of people in mental asylums suffer from guilt complexes. They feel guilty about having committed improper acts, mainly sins in a religious or ethical sense. This guilt is caused by insecurity about meeting the standards to conform with a norm. But some norms are impossible to meet—to be perfect is impossible. To be always just and good, faultless and so on is unrealistic. The majority hate themselves and have turned against their own identity. Alfred Adler said that many have their goals set too high—their own standards are impossible to reach. Maturity means having the ability to forgive oneself. This occurs naturally through insight and self-understanding. Indeed, that is what psychiatry is trying to do, but the person classified as mentally ill are closed in themselves—unshakably convinced by fixed ideas.

If one starts to think for oneself and understands the intention of the value-creating process, one should give up blaming oneself. We know that values are self-suggested and that nature is a self-perpetuating process based on imperfection.

Most individuation takes place in the unconscious. For example, the unconscious sometimes sends advice in dreams which one should heed. For example, a mechanic involved with computer-controlled machines once told me about one of his frequently recurring dreams: "I am working on this automatic machine and I get caught by the hand and the machine draws me in. I scream in desperation, 'Will somebody shut this machine off?' This is when I awake."

The message of this dream is that a frustrated individual feels drawn into a life-style that is against his identity. It is a scream for help. The machinery is merely symbolic of the way his life has become.

For practical purposes, individuals are organized in groups and must subordinate themselves to the group; this may go against their grain. The family is an organization that demands cooperation. Usually the unconscious leaves us free to lead autonomous lives. It only gets involved in serious matters, and then as symbolic suggestion.

We humans have expanded our consciousness beyond mere practical expedience and are still modifying nature to suit our needs, but this has become an artificial mode, which may be incompatible with nature. The majority of people are conservative and conform to established social values. Only a minority of secure individuals can stand alone and think for themselves. They are individuals in the truest sense of individuation. They are the creative, avant-guard of mankind, who can imagine things that most people would never think about. Formal education does not create individuals. It may help to expand in knowledge, but direct experience is a prerequisite. Apparently the main source of strength, as to whether a person can stand alone without needing the approval of other people, is based on early acquired self-worth and later acquired reason. In either case, innate suggestibility of value, which one needs to remain sane in the face of disapproval from the conservative and conformist majority, that sets standards.

Individuation is not a mysterious process. It is another step in the evolution of consciousness to understand the purpose of being—not necessarily in the details but as a general idea. It would be simple if we could tell from physical changes that a person has reached individuation, and is wiser for it. But without physical evidence, we cannot be sure. Obviously the explanation lies outside the physical level and therefore is a metaphysical question. The exploration of the human brain has proved that we only use a part of it anyway. Many persons have lived normal lives, even if sections of their brains were dead or removed because other sections took over. Obviously the brain can adjust and rearrange itself, although it may work slower afterwards. The same applies to chemical stimulations. Chemicals can produce emotional depression or mania, or can reduce these problems. Why is this only temporary? It is because the brain will soon manufacture an antidote against such help and will nullify the effects of drugs—ultimately controlling the results. If the attitude of the depressed person, however, changes to a positive one, then the body will produce a health-restoring chemical by itself to cure the illness. That should settle the question of chemical effects—a positive identity can heal mental illness if a person can sustain a positive perspective on life.

In psychology, self-understanding of the cause is called a catharsis. Insight suggests to the unconscious that there was no real cause to be ill and health is restored. Because the patient must reach this understanding by him- or herself, it takes time to convince the unconscious to give up on a fixed idea.

Presently the human brain is a hot issue. The living brain is now studied while in action, through the use of brain scanners which map electrical activity when thoughts and feelings are compared. Something interesting has been found. There is a difference between the male and female brain. In general, the electrical activity of the male brain is more concentrated to certain specific areas, and that of females ranges over a wider field of activity. The interchange of information between the two halves of the brain is more pronounced in the female. This fact prompted female researchers to think women have a superior brain. Other more cautious women have wondered if this is the case, then why are there more male intellectual leaders than women?

An explanation for this is the natural purpose of females to play an assigned role in the organization of humans from a sperm and an egg cell to a viable human being. For this purpose, a different, more organizationally oriented brain is required which could account for the difference in electrical activity. In general, electrical activity in the brain still does not explain what it is about. We can only speculate that ideas are being processed.

Females are known to be slightly more suggestible than the average male. Females also regenerate quicker from injury, both organically and mentally, because of their potential to organize cells. They also suffer more from emotional illnesses than males do, in a ratio of three to one. Obviously, their bodies are designed to undergo frequent changes during pregnancy and birth, from which they recover quicker. They also recover more quickly from mental breakdowns.

It is understood that mental illness is emotional, not to be confused with physical disability that happens to physically normal persons through injury of the brain or nervous system. This is strictly a philosophical explanation to prove that suggestibility is also the cause of mental and emotional confusion. I have neither the experience nor qualifications to deal with mentally ill persons. I know it requires patience and detailed knowledge of those who are trained in this field. This is only a generalization.

Let me recall a well-known insight of Professor Jean Martin Charcot of Paris. Paris was the city where Freud studied neuropathology, namely that mental illness is caused by fixed ideas. Charcot demonstrated the existence of fixed ideas by hypnotizing mental patients, who then became rational and behaved normal during the hypnotic trance. But as soon as he awakened them, they resumed their previous physical distortion and insane behavior. This I take as proof that fixed ideas are self-suggested. This is additional confirmation for me that the mind thinks in ideas.

After Freud returned to his medical practice in Vienna, a woman who suffered from progressive paralysis that resisted all known treatment was sent to him. I am referring to the well-known case of patient Anna X. She had developed paralysis in her legs and for no apparent reason, was unable to walk. Freud tried hypnosis and asked her subconscious to recall why she wanted to be sick. She revealed that she hated her father because he had disappointed her. Her illness began after her father had suffered a heart attack and died in a bordello. She had been called there to identify his body. She had never thought that her father would sink so low, and was deeply humiliated. After bringing Anna out of her trance, Freud explained to her that she had no reason to feel humiliated. He told her that the sex drive in normal males is a strong force, and that for some reason she was not facing this fact. Probably she had led a sheltered life, but many women recognize this weakness in men to exploit them.

The bordello is an example of direct trade. As an older man, not wanting a serious relationship, her father probably felt it is better to pay a fee, than to marry again. If he were to remarry, then die, his wealth would have gone to his new wife so perhaps he wanted to have his own daughter to have all he owned. Maybe Freud was explaining this sensible rationale to her—that her father was actually showing his concern for her. What ever Freud actually did tell this young lady is irrelevant in the details, but I am certain that he convinced her that sexuality is a fact of life. With this explanation, her disgust and anger against her father disappeared and the progressive paralysis of her legs reversed and she became able to walk again. Freud thereby made history in medicine. This was first time a paralysed person walked again.

Freud was partly misled by unexpected side effects of hypnosis. While he was assuring this woman that sex is a normal drive in man, this patient became sexually attracted towards him—which he blamed on the use of hypnosis. Freud wanted to avoid encouraging sexual attraction with his other patients, so he then switched from hypnosis to a method of relaxed conversation by letting his patients lie on a couch so they would feel comfortable and talk about themselves. Through this process which he called "psychoanalysis," he would suggest possible solutions to help them understand their problem by themselves. This, in principle, arrives at the same self-understanding, only the process takes much longer than hypnosis. From several cases involving sexually frustrated, unmarried persons, Freud was misled to think that sexuality is the unconscious drive in all organic life and built a theory based on repressed sexuality.

Sexual frustration is certainly a widespread problem among unmarried persons, but in human experience many other problems are equally important. For example, a major problem is that people do not recognize their personal value and many suffer from guilt if they had to assert themselves. All these problems can express themselves in aggression or depression. And we have learned that many forms of frustrations can affect our health and reason besides repressed sexuality.

But Freud is also well known for recognizing the channel that we have with the unconscious—dreams. And he started the use of dream interpretation and the study of dream symbols. What Freud called "libido" was organic and impulsive, but he made no connection to the material side of the world. Other psychologists, in particular Jung and Adler, expanded Freud's understanding from impulsive animal drives, to social motives that lead to a higher spiritual level. All three were medical doctors whose primary mission was to cure people so the philosophical implications escaped them.

When Freud discovered that self-understanding could cure organic paralysis, the mental basis as self-suggestion was established, confirming Mesmer. But a further success by chemistry was the "break through" that allowed access to the closed minds of seriously fixed persons who for years had been confined to asylums. Through the use of new drugs, a major step to help the mentally ill had occurred. Drugs relaxed the frustration and opened such an individual to a dialogue. When it was possible to bring them to understand their problems, this treatment was a great success. But drugs alone could not restore confused persons to lasting mental health. If the drugs are to prevent a relapse from occurring, the mind still has to come to understand the problem.

For a while it seemed that the material concept was gaining converts. But when it was discovered that the human brain could produce negative chemicals to poison itself and positive chemicals to cure itself, the mind was leading again. If convinced by self-suggestion, the brain can produce the positive chemicals to restore normal health and could do so without side effects, it could mean a lasting cure. In my understanding, chemicals are carriers of conscious ideas of their own value. This proves the chemicals are part of the universal conscious with value of themselves.

Although I had had this insight—that self-suggestion of value was the key—many years earlier I took this idea a step further. In nature, the stability of the structure of an organization depends on the belief in the value and self-worth of every entity for itself. The protection of each being against negative suggestion requires the ability to close oneself off. A closed mind is the natural protection of identity and this is how most of nature and all of the material level makes sense. Unfortunately, closing oneself reduces a human being to the condition in which life in human society is impossible. In principle, the universe is a slowly evolving value-creating process that has established temporary organizations by suggestion and by frequently fixing itself, is evolving towards individuation of consciousness.

From this explanation it becomes clear that many individuals have contributed to the understanding of nature, and that the separation and diversification is continuing as individuation. Evolution continues to head in this direction just as it has progressed from the beginning.

It eventually has become clear to many thinking persons that neurosis is a physical way to express anger while still leaving a person open to reason. However, when anger directly affects reasoning, as in mental illness, the mind is closed. The fact remains, without the restoration of meaning and value to their identity through self-suggestion, chemicals are insufficient to restore people's health.

The reason why so many believe there is no meaning, only chaos. This is because there are so many ideas that the mind is boggled unless a pattern is evident. By saying, "gestalt is destiny," Max Wertheimer and Wolfgang Koehler were implying that gestalt determines what one can become. It means that physical gestalt predetermines our life at birth. It certainly does, but one still can make something from what is given.

When one becomes fully conscious, which normally takes about twelve years, one discovers one's identity, and from that point on one assumes that identity, but does not necessarily develop independent thought. If suggestion of value is influencing the entire value-creating process of nature, one individual has only indirect influence on the process. The question is: Whose ideas are organizing the world in the larger meaning? It could be from ideas of the fields of quanta when organization begins, and later evolve in wide diversity. The process of organization is partly random and is completely unpredictable.

Since this process has been going on for aeons, it is rational to think that countless individuals have already passed through this world, which is sometimes called the "mills of the gods," and that from them other levels are already established. These levels have been established by individuals who passed through our world and have evolved to exist in a different mode, on other frequencies that we call spiritual. From extrasensory phenomena it seems obvious that spirits exist in a separate dimension above our level, a dimension that one could call the surreal. This level one can only sense and imagine. To confirm it, one has to leave the physical body. This means one would have to be dead. Most normal humans are satisfied to be alive, and can wait until they must leave this world.

Many humans in the present, as in the past, have had strange, unpleasant encounters, which made them feel uncertain about whether there is an underworld beneath our reality. This would imply three possibilities where one might expect to go after departing from this level. The less desirable would the underworld, the place of carnal, animal impulses. The second would be to return to this physical world with material values and conflicts or, the third would be go to a higher evolved, surreal dimension. At the surreal level, one would experience another mode of being with different values, but still with values and meaning.

From the general intent of the value-creating process, morality is not a main issue. But in social systems fairness is necessary to ensure peace and cooperation, but only as far as it has meaning. Ethics can be elevated and made a *cause celebre*, but like all values, morals change in time. In fact, scientific materialism has undermined the ethical standards of mankind and replaced it with consumerism and the production of material goods. Materialism rejects the idea of the survival of spirit after death as wishful thinking. Conversely, materialism is resented because it leaves no future after death. The human spirit does not die. Rather, the opposite is true—the material structures like the organic body fall apart but consciousness continues somehow. We have massive evidence of this. For now, let me return to more tangible facts, and remind us that we are physical entities. It is long known that the body adjusts to real needs. The parts and faculties that are frequently used are strengthened and the parts little used are reduced in strength. This is strictly adjustment by experienced value. That explains why the brain opens additional circuits if demand exists, or reduces the circuits if parts of it are unused. This is evolution by suggestion of value in principle.

For example, if one is excited by an idea, it can cause the body to develop chemicals that result in physical change, such as the expansion of consciousness and individuation. One may have influence on the direction in which one evolves by the life-style and suggestions of value that one pursues. This is conceivable because the brain would add the circuits to that avail, as we know it does if we use it.

The chemicals that are produced by the brain are designed to maintain normal function, but they can be influenced either to destroy or improve the brain, according what the individual has concluded the value of life is. It has also been proven that mind and organic needs affect each other. The sex drive, for example can be explained as caused by hormones. Hormones also produce sexual fantasies. Or, in reverse, sexual fantasies can cause hormones to shoot through the body, particular when circumstances present themselves for fulfillment of the fantasy. In either case, organic or mental, it is suggestion of value, because the body also projects ideas into the brain, even in pictorial dreams of action. The implementation into physically manifestation is a detail which we do not understand in depth because the unconscious deems it unnecessary to burden us with details.

Let's return to the subject of organization. As I said before, if chemicals can create emotions, than emotions should also create chemicals in turn. Although we now understand this, it takes a while for the body to react to producing chemicals to counteract an invasion of bacteria. In the meantime, one is in danger of dying. During this period, many general drugs may fill the gap—that is where medical science proves helpful. Drugs are necessary to help until the brain develops a specific chemical to heal itself and the body takes over. But the body will not achieve a real cure from the general drugs without the wish to live.

We take positive reaction for granted. What is of concern is the negative reaction that disturbs health. The mind can indirectly paralyze muscles, block organs and develop allergies that are totally illogical in chemical terms as we know them, but make sense emotionally.

Allergy is basically a response of the body against unwanted influences that goes against the self-image. It is expressed on the cell level, by causing swellings, usually in joints. If the swelling occurs in the windpipe one can suffocate in minutes. Allergies defy scientific explanations but make sense in psychological terms. This is where medical science ends and psychology begins. If the cause is emotional and self-suggested, allergies are emotional reactions. Because many illnesses happen unexpectedly, as most illness does, we understand that the immune system holds the invasion up as long as it can, until finally it is overwhelmed and breaks down. But what can we do against sudden illness? Is it simply a weak immune system or is the immune system turned off intentionally?

I know a case, in which a married couple had their first child, a charming girl. They adored her and lavished her with attention, but this attention changed when the mother gave birth to a boy and he became the center of attention. All seemed to be going well, until the girl developed eczema on her skin for which ointments were prescribed. For a few days the medicine helped to suppress the symptoms, but soon the eczema spread. The doctor concluded it was food allergy and ordered no more foods that contained sugar. This diet slowed the spread of the eczema, but soon it returned. By the time the girl was four years old the parents' affection had become equally shared between their children. The little girl also found other playmates, and soon the eczema disappeared on its own.

Assuming we all have problems in life, some people are able to take them in stride, others make more out of them. If the situation offers no hope of improvement, instead of venting anger freely, some persons will repress their feelings and drive anger in the subconscious where it is transformed into a variety of physical ailments. These ailments do not respond to normal medication, but often will respond to suggestion in an a hypnotic trance. But there is a danger, While suggesting a physical problem away, repressed anger could be transformed into a mental problem if the cause was not eliminated. But it is the search for a cause that can revive repressed tragic events that should have been left buried. Discovered, they could make the problem worse. Many hypnotists therefore leave the details of the cause undisturbed and stay with general reasons and to avoid further involvement.

Another approach to understanding the unconscious is from dreams which can tell about inner problems. Freud had called dreams the royal road to the unconscious. Dreams are helpful indications for the treatment of physical and mental illness, by revealing their underlying causes. Recurring dreams mean that the unconscious wants to tell something. Fear-inspiring dreams, called hallucinations, can literally scare people out of their wits and drive the insane even more insane.

Jung, who once was a devoted pupil of Freud, had begun to search deeper into the unconscious, and found that the unconscious had also a spiritual dimension, something that Freud rejected. As director for the Mental Institutions in Switzerland, Jung had a different perspective of insanity and noticed that the majority of the seriously ill patients suffered from grand delusions of importance and of religious delusions caused from guilt. Jung assumed that the unconscious is individually different. This theory contradicts Freud who insisted on sexual and organic motives. Jung thereby broadened the perception of the unconscious and added a spiritual dimension. After studying dream symbols, Jung concluded that a human related "collective unconscious" must be assumed because the dream symbols and dreams in general have human meaning. I agree with the idea of an individualistic unconscious as collection of human values because other species have different values.

Besides Freud and Jung, another psychologist, Alfred Adler, also made sense. He saw social relations as the main reason for emotional problems. Unrealistically high expectations result in feelings of inferiority later in life, such as being a loser, and then hating oneself. If one is realistic about oneself, one will avoid setting goals that are too high to reach. In my personal opinion, life is an interesting experience, not a competition between individuals.

Freud, Jung and Adler were early pioneers in psychology. They knew each other and disagreed on details, but their principles are right if put together. I consider their ideas as a pattern of the cosmic psyche which repeats itself in many modes of consciousness. In the human experience of reality there is the physical level of tangible reality (Adler), and below there is the unconscious organic libido (Freud), and above, the higher collective unconscious (Jung). Each human has the physical experience as a main identity, but is also aware of the organic sexual libido, as sex drive, and in dreams is also reminded of the connection to a higher collective unconscious. Each person has these three levels, but in different degrees. Some have a strong libido, others are more practical and expedient, and some persons are high-minded and spiritually interested. This is also reflected in their actions and dreams. What they dream is what their unconscious prefers.

This is a human experience of reality which is as old as time. Early man thought that below the living exists the underworld, of the dead and demonic forces. Above them lived the unseen gods. The gods were unpredictable, but could help man if one stayed in good standing by making offerings and sacrifices and by begging for help and mercy. Later in history there was a shift to monotheism. The demonic underworld was called Hell and the many gods were combined as one God with a moral character. Morality was the deciding reason whether one will go down to hell or up to God who lives in heaven. An example of monotheism is the Judeo-Christian model, as given by God to Moses in form of divine laws. According to this model, life is a moral test, sorting people by good and evil. If one lives a perfect life, after death one comes to heaven near to God. If one totally disobeys all moral laws one is cast into Hell. And for the all the other souls who led imperfect lives, they must remain in purgatory until they mature. Let us assume this permanently imperfect world is a purgatory in which one can evolve to a higher consciousness, except that God is a higher dimension of consciousness. This in principle is a parallel to the process of individuation. It is also similar to the Asian model of the Hindu and Buddhist concept of karma. The individual karma matures from each life experience. One is reborn into life again and again until one reaches understanding of the meaning of being and is released from reincarnation to join Brahma or go to Nirvana. This is a basic pattern which is common to the unconscious of mankind and may indicated that one matures in life according to the choices one makes. If one wanted to go into the details of each model, then the differences in values and modes obscure the pattern and lead to estrangement and conflict. This natural development of differentiation may be intentional, namely to make the world an interesting experience, and is intentionally imperfect to remain incomplete as continuum of imperfection.

To determine whether this deduction repeats itself in the real experience of life it must be compatible with the conclusions of psychology. Jung analyzed the dreams and hallucinations of his patients and concluded that from the dreams' content, a spiritual dimension beyond death must be assumed. He called it the collective unconscious—a pool of the collective memories of mankind. This unconscious pool of memories can be a spiritual dimension, just not of a holy spirit. It is a collection of normal human experiences. According to Jung, mental illness is the dark side. The negative aspect of the unconscious leads to the disintegration of the individual identity. The positive side evolves more slowly and in holding steps, securing behind itself what it has created. Destruction is much faster than evolving by growth. Negative expectations can reduce a human being to the mental level of animal or vegetable experience which is confirmed by statements of such confused persons that time is irrelevant.

This process of reduction to a lower idea of existence has many parallels on other levels of nature. For instance, on the nuclear level, I noticed a classical case of reduction that compares to human depression and insanity. The naturally occurring uranium isotope, which is slowly disintegrating, is on a negative trend, like humans feel when they have lost interest in life. Proof of this is presented experimentally if one encloses a quantity of refined, decaying uranium 235 in a box of pure graphite. Then, the existing radiation is prevented from dispersing to the outside, and reflected back on itself. This causes a build up of a critical mass of repressed energy that then attacks itself, similar to mentally disturbed persons who turn on themselves. In nuclear fission, the critical mass can overheat and explode with violent force. By comparison, in atoms it becomes a nuclear bomb, in humans is a berserk murderer shooting people at random.

The pattern of isolation and turning on itself is similar in atoms as it is in mental illness of humans and animals. At first there is the negative disposition, which in humans is a skeptical personality. If one is disagreeable, one will be avoided. This isolates such persons and their negative thoughts then turn inward on themselves. These are usually the quiet people that hold their anger in. That is why other people are surprised when they go off like a bomb, blaming others for their anger and discontent.

With some imagination and a little knowledge of astronomy one can see a similar emotional pattern in the macrocosmos. Galaxies die and are born, creating their own emotional energy, and in the process reorganize themselves to form new worlds of meaning in which beings like ourselves evolve. This is certainly metaphysical speculation compared to the "Big Bang" theory, which is a mechanical deduction. In general, there is far less proof to the Big Bang theory, except that it is mechanical and as such is attractive to physics as a predictable model of the universe. And, the mechanical theory allows mathematics to be used.

From the psychological point of view, a much larger pattern of an emotional universe can be deducted then from a mechanical theory. The psychological model fits into the single principle of reality by which individual humans live. It makes more sense to explain the purpose of life an interesting experience than any lifeless mechanical formula ever will.

The idea that higher-evolved beings, considered unpredictable gods, could exist on another dimension, is even older than all other beliefs. In fact, this is an empirical deduction of the nature of reality. Directly perceived, we live in the physical world, but indirectly, behind the tangible world there is the metaphysical principle of innate suggestibility, which can explain the whole as harmonious system to create meaning.

The hallucination of the mental patient indicates that without a solid identity, one is defenseless against unwanted ideas and images that want to influence us. Obviously, the cause for mental illness is a weak, self-suggested image of worth.

Because we humans have become specialized into professions, skills and other indoctrinations, even our perspectives of the outer world are individually different. In philosophy, the idea of "creative conflict" as the basic drive in nature is a logical doctrine. But it cannot be that conflict itself creates, but suggestions of value which often coincide with conflict but do not depend on it are the creative force.

Jung extended individuation beyond the natural process of maturation to the transcendental or supernatural level. He differed from others by his use of extrasensory phenomena and dream experiences as indicators of persons recovering from an injured identity. Jung believed that his task, as a psychiatrist, was to make fractured persons whole again. He intended to become a therapist who healed people. Although he had wide knowledge, and influenced metaphysics greatly, he was still impressed by experimental science.

The humanistic psychology of Abraham H. Maslow also has relevance to individuation. Maslow studied the type of person he calls the "self actualized" individual. This is someone who is open to new ideas, with superior perception of reality—a person who is spontaneous and creative. Self-actualized persons accept themselves and respect others, focusing their thoughts on problems outside of themselves, instead of feeling guilty or incompetent. It seems that individuation and self-actualizing are connected in principle. The process of individuation is certainly on a "psychic level." It is also called "trans-personal or idiographic" psychology, which is not actually transcendental, because the mystical aspect is avoided by Maslow and others. This school also interprets dreams but leaves spiritual meanings out. This I think is a deliberate ignoring of important clues. Yet, Maslow also thinks that frustration and repression do not inspire creativity. Creative conflict is only an indirect cause of individuation or evolution. Also, competition is not typical of creative persons. They unfold in freedom and unforced lifestyle. In fact, evolutionary advance is caused by avoiding conflict and searching for alternatives—thereby new things are created.

Destruction clears away old structures and creates opportunity for new ideas to organize. One has to see the larger picture of the cosmos and humanistic psychology with Maslow who has rejected any spiritual dimension above the physical level. But in the nightly dreams of normal humans, we can confirm a personal connection to the spiritual dimension. The persisting rejection of this other dimension has been blocking western philosophy and psychology since Galileo and Newton mechanized nature, and has even partly influenced psychology. Agreed, it is a complex issue that would allow transcendental connections to exist, but even psychologists are refraining from unpopular conclusions, because the establishment is dominated by scientific materialism.

Among psychologists and psychiatrists, such as Wilson Van Dusen, the opinion has gained ground that some dreams of their patients actually are clairvoyant and contain meaningful messages for the patient and also forecast future events. But when connected to mental illness, these are difficult to separate from confused experiences. If these confused individuals only could calm their "deep fear," they might understand that they had transcended their separate identity and need not think that everything is insane. But a rigid mind and fixed values prevents this insight, because they are convinced that this cannot be. They are usually "paralyzed by fear," and therefore unable to see any meaning in their experience, other then a threat to their sanity. They need to understand that this confusing experience is a break-in from other dimensions. It is a passing interface with the timeless consciousness—a consciousness of which the ancient seers said "it never is not."

This means the mentally ill are not necessarily insane. They are only open to unwanted influence for which they are unprepared, which does not automatically endanger their life and sanity. However, a disclaimer is necessary. The extreme suggestibility of frustrated individuals makes all conclusions uncertain. Essentially, every life is a separate experience and should stand for itself. This means one cannot draw conclusions from the experiences of individuals about the whole of the surreal dimension. Obviously the next level is on a different mode of reality. Jung thought that the unconscious of humans is on an individual mode and the general process is towards individuation.

Rational persons know that their dreams have to come from somewhere. If from the unconscious, what is the unconscious? In his professional experience with mentally disturbed persons, Jung came to the conclusion that a "collective unconscious" stands behind our apparent individuality and will take over, if the individual consciousness does not perform satisfactory. The unconscious indirectly reveals thereby, that it intents for each individual to assert itself, its own value. If the individual fails to satisfy this basic expectation, the unconscious steps in and asserts value for it, and thereby acts over the individual's rational judgment. This can explain the irrational behavior we call insane—the unconscious misunderstands the outer world, wants to assert itself and consequently is deliberately overriding of the realistic judgement of conscious reason which causes insane behavior. Because the unconscious is fed up with weakness and mad about losing out, this is an example what happens when blind will takes over. The result is that angry persons who are out of touch with real life must be restrained, even forced into a straightjacket in order to protect other people and themselves from their fixed ideas. Inner conflicts can also be traced to guilt-complexes arising between loyalty to outdated values and reason that is demanding change. One feels like a traitor to no longer trust old values. Often it becomes a serious crisis that one has to go through to evolve further, to free oneself from instilled ideas, such as religion, that one can no longer accept. Jung concluded that in the collective unconscious could also be the memory of the cultural values of a people, even of mankind. This memory reveals itself in the content of certain unusual dreams that contain mythological symbols and images, many of which the dreamer has never known before in this life. Therefore, one must conclude that dreams transcend the experience of a single life. I agreed that most dreams of normal people are concerned with daily problems and suggest solutions for existing conflicts in this life. Only occasionally does one experience a "grand dream." However, such exceptional dreams differ from the usual personal concerns and are unrelated to the daily experience which the unconscious rephrases in dreams. From this, I conclude the collective unconscious is the other half that is always available, but the higher mode is a further development of individuation. It is perhaps a higher advanced mode of consciousness, when the conscious and unconscious self become "one" as a new experiences on another level.

Great dreams happen occasionally and are marked by the presence of meaningful organization and clarity and the fact that one remembers them. Obviously these dreams are meant to inform us of events in general that lie ahead. Because details are sketchy and symbols are used means that the experience is still ours to have. This indicates that existence is meant to be an interesting experience, should not be diminished by giving it away. This is of paramount concern—the meaning of life.

That we all dream was scientifically established by experiments in dream laboratories. We apparently need dreams as the exchange between conscious and subconscious levels for our mental stability. Much has been learned about the physical side of life, yet it is the sub- and unconscious side, the transcendental aspect, that we know little about. The interpretation of dreams is one opening which reveals important facts about the unconscious. Human life is a balance between opposites in values; judgement keeps to a middle ground between the extremes to avoid unnecessary conflict.

From a philosophical perspective, materialism only means that the world is made of matter, and idealism is that the world comes from ideas. A middle position is that the world is made of matter organized by ideas. This is not exactly true, because reality is strictly a wave phenomenon and matter an illusion, but it is near enough for the level on which we perceive waves as solid. Nature is obviously intent on maintaining the dynamics necessary to make life interesting, and through the continuous conflict of values, change is guaranteed. Going back to Jung, he also used words like ego and alter-ego, and the moral expression, "persona and shadow." The shadow is the dark, the repressed organic and occasionally violent side while the persona is the exterior, or the false mask of proper character that we present to others. To maintain the stability within, we must balance both sides to remain whole. Jung also recognizes the creative duality between female and male values which serve to accelerate the process of individuation. It is an established fact that, in order to remain normal, we require stimulation and opposition, but moderation is needed to maintain a balanced marriage.

Even if the goal of psychotherapy is to reduce conflict, life can never be totally free of all problems, otherwise it would become boring. One would have to withdraw to a fixed and autistic existence, such as the beings on the material level do. They are mostly self-contained and appear to be dead. I agree that even the material level cannot be totally unconscious because subconscious channels exist which connect everything emotionally. Other dimensions and levels have some consciousness and influence on our lives.

Let us return to the subject of dreaming. While many individuals may claim that their dreams never reach beyond personal interests and usually make no sense, this does not exclude the possibility that many other persons can have clear dreams of extrasensory content that are far from ordinary experience. Many specialists compare our brains to information-processing computers with memory banks, but the comparison does not go far enough. Besides receiving information through our senses, we also send out messages as well. Some of the exchanges may happen without us being aware of them. Nevertheless, within our species, suggestion works unconsciously and indirectly, namely as self-suggestion. Even if we adopt ideas from others, they are first screened for value and modified before they are accepted as useful. Because suggestion, in order to be effective, must be converted into "self-suggestion" otherwise it is rejected. The mental process of individuation has been described by the great writers in the literature of mankind. These great writers express their innermost feelings of their own psyche and we are able to identify with them because of their talent.

Literature, like nature, uses the conflict of opposites to create the interesting dynamic. Without conflict, nothing of meaning would happen—not in theater, nor in a book, and nor in life. Although heroism seems to have symbolic value, originating in the mythical past of mankind, it is outdated. Now people admire greatness as excellence in the arts and whatever furthers human self-understanding and advancement. Perhaps we have reached a point in the process of individuation when man understands the principle that inspires being, namely through the suggestion of value which evolves ideas to organize into manifestations of meaning.

This question is still unresolved: Is the next dimension individual or collective. Literature and religion, particularly spiritualism, claims that our souls survive death. This implies that we remain individual entities. They support this belief through personal massages from the departed spirits who are on the next level. The messages come through psychic mediums. Yet, psychology, through the interpretation of dream content, must assume that many symbols are general and collective. This implies the existence of an impersonal, collective mind-field, as if our side is individual and the collective is its opposite. This suggests what could happen to the soul after it leaves the body—it could become a part of the collective consciousness with the ego in a temporary limbo unless it chooses to manifest itself again. But it could go to a different level. Nobody knows.

The answer is already given. We do not need to know what will happen since uncertainty is a necessary element to keep the world interesting. Because the next dimension is uncertain, its occasional interference in our existence makes it real. Therefore, I prefer to call the next dimension the surreal. It is not the same as surrealism in the arts, but a level above our reality.

Most philosophy deals with trying to understand reality and to live in harmony with organic nature. The outer pattern indicates that individuation divides us, but the unconscious reveals that we are internally connected. This general pattern of outer separation and inner connection repeats itself infinitely. This is almost like the planets, which are kept apart by distance, but by identity are connected from becoming lost into space. We are kept separated at a distance from each other individually, but unconsciously are part of a meaningful value-creating process with human meaning.

From a philosophical perspective, the structure of known levels below us suggests the existence of more levels above us. The existence of at least one higher level above mankind is suggested by extrasensory experiences. A need for another dimension was always there as anticipated in the branch of philosophy called "metaphysics." In fact, most philosophers think that all philosophy leads to metaphysics and the solution to the understanding of reality lies outside the physical level.

During speculation about the Universal Principle, the question about the goal of individuation comes up, suggesting we are being prepared for a next level, for a more refined mode of being. Assuming we are only passing through this world and not meant to stay long, we could become enamored with physical life and return if we wanted. It could be that on the next level a higher integrity and maturity is required. It could be that the value-creating process is aiming to achieve a higher level of sensitivity through expanded consciousness.

Contrary to this, many well-meaning persons are pointing to the process of separation and differentiation as the cause of all evil in the world. They are preaching that we should break down the separations and divisions among us, to become brothers and sisters, and one with the Universal Mind. However, without separation and differentiation, without distinction of meaning, there are no separate individuals or particular values, only equality in non-being.

The normal meaning of life is to follow evolution forward and we expect to transcend to a higher level of consciousness. All else is misleading because scientific materialism is a misconception that misses consciousness as essence of being. Nature intends us to participate in the value-creating process, which may involve pleasure and pain, but with foresight, pain can be reduced. Pure consciousness existing in the void may have created itself worlds of meaning to escape boredom, and now has achieved self-understanding in the process of organization of ideas. Individuation means personal evolution inside the larger evolution, or processes evolving within processes. In principle, everything repeats itself, but in detail, everything is new and diversity increases by further division and reorganization. Obviously, this is metaphysical speculation. The physical model is insufficient to explain conscious beings. It is a fact that the general process shows gradual improvement from quantity to quality, typical of a value-creating process toward individuation.

10

Transcendence

WITHOUT SEPARATION AND THE DISTINCTION of values there would be nothing. What exists has evolved slowly by differentiation. To transcend the separation is counterproductive to the value-creating process. This makes transcendence a controversial subject. For a long time humans have tried to transcend, but only with limited success. It is possible to transcend in thought, and transcendence is quite common in dreams and deliberate in meditation. But to transcend physical barriers is difficult and has been mainly achieved as a negative, involuntary act. This has happened when frustrated individuals have seemed to awakened the sympathy of material substances, which acted on their behalf. This is demonstrated for us in psychokinesis with the cases in Rosenheim and Tucson. Alchemists from the Middle Ages believed that sympathy could be evoked by giving praise and making wishes. Indirectly it means that they believed that materials are suggestible. Among the alchemists in Prague were many intelligent persons who had as much knowledge of the Greek philosophers as we have now. When some of the theoretical physicists became attracted to Buddhism, we see that this is a return, at least in principle to transcendental idealism.

During meditation, monks transcend to the essence of consciousness, but I personally think this is the direction back to nothingness. On the contrary, I want to progress to interesting experiences in life. We know that parapsychology, with the help of physics tries to influence nuclear energy. But whether practical results come from it remains to be seen. I am content to wait to see if the unconscious wants to talk to me. I enjoy being part of the psychodynamic universe as it evolves, knowing that we are only small replicas of the world spirit which knows more than all of us, and leaves us alone to experience meaning.

In dreams, the mind exchanges information that transcends normal experience pointing to the unconscious as a source. Mankind has been dreaming for a long time. Why should we ignore this connection if it helps us get out off the present materialistic stagnation? Transcendental dreams and visions helped us to break through the barrier that separate the individual consciousness from the collective unconscious. Transcendence is the last avenue of knowledge left to be explored. We know that certainty is elusive—this leaves uncertain speculation as still better than nothing. To call my speculation uncertain may not be completely correct. I make frequent comparisons to other levels to ensure that my thoughts are compatible with already known facts. One such fact is that we all are indirectly connected to a larger unconscious. It has the combined knowledge of mankind and the insights of all philosophers and may exist in a timeless state, knowing something of what we desire but is not readily available unless seriously pursued. In addition, we have the advantage of having large pools of written records of experimental data and the collected ideas of many centuries available for comparison with extrasensory information.

By necessity, some form of transcendence connects us through thought waves. These are comparable to electronic waves which travel through space and carry information. The difference is that thought waves are of a another mode from electronic waves, undetectable by instruments. This makes sense as a means of protecting us from each other.

The human mind can apparently connect and receive information—even participate in the experiences of others—but only when off guard or asleep. Transcendence of consciousness happens on a mental wavelength on which the brain can receive as well as send experiences on unconscious channels mostly of current relevance during sleep while often disregarding time and space.

One important question about which man has always has been uncertain, is this: Is there a source or sender of dreams and for what purpose? We know that television requires stations to send out meaningful programs, but who sends out dreams and why? Let me just say, television sends out messages as information and mostly entertainment. Could it be similar?

Cautious professionals want to tell us that our own unconscious creates all dreams. That may explain dreams of ordinary experiences that concern our personal life. But what explains dreams containing information about events that are happening far away, and before their time, which later are confirmed by actual events? This is a question we must seriously address. To avoid it, even deny it, is dishonest. If one knows that this can happen, one must face it and include it as real.

Since most dreams have meaning, it is important to learn to understand the symbols in order to comprehend the meaning. Understanding has always been a problem because it requires interest and some knowledge of the collected data of other humans who wrote down their dreams and compared them to what followed in the real world in order to make sense of the meaning. Much information has been gathered in dream books dating back to ancient Babylon and Egypt, but this information from another time has been greatly expanded through the records kept by psychologists and psychiatrists in the twentieth century. Their records are mostly of disturbed persons whose dreams are less significant for philosophy but are still helpful.

Jung broadens the range of dream symbols to include spiritual images. Since then many others have contributed greatly to additional understanding of dreams. In particular, there are dream laboratories where normal persons are awakened when rapid-eye movement indicates dreaming. It is long confirmed that everyone dreams, but we mostly forget the dreams when we awaken. This is good, because one does not need more confusion in life by letting the unconscious enter in. The question which some poets have raised, whether life is a dream, is easy to answer. This life is real because it is continuous, but dreams are only fragments in symbolic language, suggesting ideas.

If dreams suggest ideas of value, then there is something wrong with the value that dreams suggest. They are mostly confusing symbols and cannot be taken literally. But if one thinks of dreams as a theater of the imagination, the "theater" reveals something about the unconscious, namely that it uses dramatization and exaggeration to tell us something that is far less dramatic. For example, before she got married, a young woman dreamed she stood in a long line of women, all patiently waiting to get their heads cut off by a guillotine. She awoke just before it happened to her, thinking what nonsense this dream is. But the dream did not mean death. It was a dramatization that in marriage she could not expect to have her own will. That other women patiently accepted this fate, made it unemotional. Dramatization is one aspect. The second aspect is the range of the emotional impact. This is important to know how serious a dream is. The dream symbols are interchangeable, often meaning the direct opposite from what may take place in the real world. The execution in the dream meant a change of life-style, symbolically preparing the young woman for what marriage means. A funeral could mean a wedding, or the opposite of a wedding could be a funeral. It simply means a big change is going to happen. I ask, does this kind of thinking consider death merely a change of experience?

Emotional dreams are more serious. For example, if a person dreams that the world comes to an end, sees the sun disappear, the moon and stars fall from the sky, and is emotionally shaken up and convinced that judgement day is near, then one can take this as indication of a mental breakdown. To be absolutely convinced this will happen is a fixed idea with emotional impact. This means that for this individual only, the world has come to an end. Although such persons will insist that we all should prepare for the end, it is only their world that has collapsed. They become emotionally incapacitated by fear and are unable to function rational. Sometimes a person with a confused identity thinks he is a prophet. In fact, many prophets may only be suffering from an emotional depression, which causes negative and catastrophic expectations to dominate their thinking and they feel compelled to warn us all from a coming doom. Even for normal persons, dreams are confusing because in dreams we transcend space and time. Life and death are interchangeable for the unconscious, like opposites that belong to the meaning of being. The place where dreams come from could exist outside of the physical range of the four dimensional space-time continuum, on another mode of being, with different values.

Dreams usually are without religious content, without moral or ethical feelings, except if persons are extremely religious or guilt ridden. In fact, most dreams treat dramatic events rather casually—as if they were theater, where a change of scene happens frequently in order to present a message. Changes of scene and sudden transformations strike us as unreal, because things on our level evolve gradually and scenes do not suddenly change, nor do individuals and objects transform instantly. From this one could conclude that the meaning of dreams is to suggest ideas particularly through active participation, which is the best form of learning.

Dreams also reflect our inner values and have distinctly individual orientation that reach beyond the manifest world in the realm of ideas. Individuality is a major factor for the interpretation of dreams and symbols. Although I shall try to explain dreams by giving examples, they are only my own interpretation or what a person confides in me. Many dreams are similar in principle, but are distinctly individual in meaning.

This reminds me of philosophers such as Socrates, who discussed the two-level world—ours and that of eternal ideas. And of the *Rig Veda* claiming: "We never are not." Together these separate opinions could mean we exist as individuals in a dual experience in a temporal and eternal consciousness, to live through interesting ideas that suggest meaning and value.

Now in the age of the psychology of the unconscious, when the question of transcendence has to be addressed, the unconscious reveals that we all are remotely connected with the eternal consciousness. One begins to think that the insight of Lao tzu, that the world is the "one in the many" makes sense. It is logical that the world is created of opposites that give meaning to each other. The opposites are necessary to create change, telling us that this reality is a value-creating process, which creates itself by creating pairs of opposites.

But we cannot suggest values without realism. Realism must be enforced—that is where pain and pleasure make themselves known. Only if one contests the reality, will it need enforcement. If suggestions of danger are enough, suffering is avoidable. If we foresee and respect what must happen, one can avoid the consequences and live a reasonably long and healthy life.

If one could ignore the existence of the unconscious and the fact of dreaming, then such metaphysical speculation would be nonsense. But with modern psychology this dimension has become even more difficult to deny. Still, many people call transcendence deluded. Yet, the world is continuing to dream of interesting experiences, which intelligent humans seek to explain in general terms. It is reasonable to conclude that we should eventually understand our unconscious, particularly if one admits that occasionally ideas come to our minds through transcendence while we are resting or sleeping. One way to explain this is according to the three stages mentioned earlier: First, by saturating the mind with a problem; second by giving it time to incubate, and third, by receiving symbolic answers in a dream or while resting with a relaxed mind. In principle this is how transcendence works. Waiting is perhaps not for everyone; sometimes the ideas never come. One does not have to be aware of it, as long as one becomes deeply absorbed, but then it can be explained as logical deduction from saturation.

Survival of consciousness

In ancient Tibet, transcendental monks compiled a book about the experience of dying, and what happens after death in the next dimension. The book, called the *Bardo Thetol,* says that each departing soul seeks its own level, a level where the soul is familiar and prefers to be reborn. The monks also mention different colors of bright light as attraction, to guide souls to where they fit in. There are similar accounts of "near death" experiences in our time from a medical doctor, Raymond Moody, in a book called *Life after Life.* He writes about reports of clinically dead persons who were attracted to a bright light. This implied innate suggestibility to be attracted to familiar surroundings confirming orientation by suggestibility to value, and thereby we prepare our future in the lifestyle we pursue now.

Most of the ideas of the immortality of consciousness and rebirth are old, new is that the meaning of dream symbols point to similar conclusions—even if the details are different, and gods were given credit. It is probably true that many ancient philosophers and prophets derived their ideas from transcendental experiences, but in their own times knowledge of the extent of the physical world was limited to the human experience, and their concepts were of a flat world. Most prophets, except the philosophers, claimed their ideas came from higher sources—a position which they unbendingly defended. Regardless of whether their knowledge of the physical world was limited, compared to now, these early thinkers were catalysts for social change. Their ideas sparked debate over differences in values that frequently resulted in violent disagreements—values which we today consider unprovable and relative in detail. Yet it must be admitted that the mystical world-view was relevant to the condition of their time. Today we understand revelation more as symbolic, dramatized and subject to further interpretation. Nothing but a general principle can be absolutely true in detail in a constantly evolving world.

Intuitive insight, in our century, is called extrasensory perception. Because it only occasionally transcends the physical limits of visible matter, space and time, it is therefore referred to as the gift that only a few individuals have. As we found out, even these humans are only selectively gifted—usually in one field. Some are limited to the emotional level and can feel emotions even in material objects. When others see images and events take place before they happen, it is called clairvoyance. And then there are individuals who get ideas about nature or in health. The gift to see connections has survived the skepticism in western education, and is still accepted as gift among the occasional scientist and philosopher when larger theories are the result. Because such extraordinary ability is beyond education it belongs to transcendence.

If we accept the idea that we occasionally receive suggestions from an external source, I think that the suggestions are neither from a divine source nor outside of the value-creating process, but from a more advanced level of consciousness—ahead of the level of evolution that mankind is on. Apparently a certain empathy for our trials and tribulations exists on a higher level, implying that these beings have understanding of human values and probably belonged to the human species at one time. While their concern indicates understanding and sympathy, these beings refrain from direct interference into individual life experiences, as if we are meant to learn by ourselves to reach maturity.

At this point, the question could be asked: Is this other level a collective dimension or do individual beings dwell there? As far as images from dreams are concerned, it could be both. Spiritual mediums claim to sense or even see individuals as they once were in our world. There could be many levels and stages, however, where collective and individual consciousness are just opposite ends of consciousness itself. In fact, a similar uncertainty is also encountered in physics. It is now also unclear whether waves or particles are basic in nuclear energy. One could speak of wave-particle or wavehicles. Yet, if suggestibility to value is innate to quanta, then excitement in quanta fields could be the ultimate cause of waves that organize consciousness to evolve and create meaning. The question whether life is a dream was raised many times before, and one could answer simply that wherever consciousness becomes conscious of that is real at that point. Dreams are mostly experienced in a picture language and later included spoken words, even written pages and numbers. These later symbols indicated the adoption of new values, which means that human symbols are changing also. For example, a friend said to me, "Because you believe something, actually means that you do not know." I agreed because that is all that can be. Even persons who claim to know are only assuming they do. The problem with words like believing and knowing is that it depends on what you mean. Speaking for myself, I like uncertainty, because it makes life interesting and if knowing means certainty, I am skeptical that we know the absolute truth.

Therefore, the examples of dreams I can mention are those reported by reliable persons, who I believed tell the truth. Prophets, who are often called seers, seem to know generalities rather than details. For example, a story written in the Old Testament, dating back to ancient Egypt, says that the Jewish prophet, Joseph, was asked to interpreted a dream of the pharaoh, the king of Egypt. The king repeatedly dreamed of seven fat cows being devoured by seven starved cows. The prophet told the king that the dream mean that after seven years of plenty, seven years of starvation would follow. One could argue what have cows in common with years? That is why one needs an open mind. The symbols of "fat cows" fitted into the ancient values of prosperity, three thousand years ago when cows had value.

In the twentieth century our dreams have modern symbols for value, as for instance this dream that a friend told me: "I was in a hydroelectric power station, where a row of turbines stood there idle, failing to generate power. The problem was a breakdown in the system somewhere which the engineers had not figured out yet. The operator in attendance said, 'Something is wrong with the computer, we have to sort this out.' I walked to the window overlooking the dam and saw that instead of water, there was a large mud flat that had drifted in and was blocking the intake. I called the operators over and said, "Do you never look out of the window?" Then I awoke. What a difference in values—electrical energy instead of fat cows are among symbols of our electrical age on which the prosperity of a nation depends. The meaning in this case is that something much more "natural" is implied than malfunction in a mechanical computer. Although the symbols are modern and technology is used to suggest value. But for the ordinary life of the dreamer, energy is derived from food. It could mean that the system breakdown was the result of eating "mud" in the food. This is a current problem we are being poisoned by the processing that extends the life of food with preservatives. I am aware that this is an alternative to being poisoned by germs in a slower way.

To the individual who had the dream, the dream suggests that he is eating the wrong food which is the is the reason for his low energy and bad health. The dream uses grand ideas, such as power stations and dams, to tell a simple message. In fact this young man had switched to a vegetarian diet because of allergies and still had low energy, not realizing that even vegetables are fertilized with chemicals.

Dream symbols frequently dramatize problems out of proportion, which means one cannot take dreams literally, but they still have meaning if one scales everything down to our level. Since the average individual becomes aware of transcendence only in a crisis, or preceding a crisis as preparation for an unusual event, transcendence remains outside of experimental proof.

Why are only some persons paralyzed by fear in dreams about a crisis, when most remain calm and detached? Is the cause insecurity or low self-worth? This is a good point: Many mentally ill persons are highly intelligent, but have a weak self-image. Their self-image is uncertain of its values, even if assertiveness is outwardly displayed. This inner fear has been called existential angst—a fear which seems in part to originate from rigorous training in the educational system that breaks the self-confidence of the individual. The nihilism of the scientific world view is a common syndrome among intellectuals. Angst is mainly caused by frustration of their drive—by high demands set in childhood. The belief in a mechanical universe is attractive to persons with a nihilistic perspective of life, whose spirit was reduced to mechanical thinking by mathematics.

Many scientifically oriented individuals are opposed to a transcendental concept of reality and uncertainty. The mere acknowledgment of intentionality would imply a purpose in nature. They will not accept this because it would demand that the material concept of the previous three hundred years should be abandoned, and a return to magic and animism might occur. But this concern is unfounded because we have already much understanding of the extrasensory phenomena and could never return to a belief in magic. Even the alchemists have improved their perspective. But the other extreme, to reject consciousness and ideas and place trust in experimental science has mislead many educated persons to disregard reason.

The interpretation of dreams and visions has changed since the days of old Babylon and ancient Egypt when astrology began. Psychologists since Freud and Jung have kept records of dreams with the progress of their patients and discovered more details about the meaning of symbols in connections to reality. We now possess a fair collection of useful symbols and have agreement about what they could mean.

Ever since I became aware that reality is a psychic phenomenon, I have turned my attention to psychology and psychiatry to acquire at least a basic knowledge of the art of dream interpretation. This I did by reading the books of Freud, Steckel and Jung. Obviously, I read ideas into them, which might not be there, because I have my own mind. Since I am more interested in the philosophical side and less in the methods of healing, my views have little relevance for curing mental illness. That should be left to professionals.

Many readers will know enough about the subject of dream interpretation, to recognize that it is from a different perspective that I see it. Nevertheless, I want to make it clear that my view is philosophical speculation. For example, it is agreed that a dream about a house means the physical body in which the soul temporarily lives. The broken door of a house in need of major repair means the house has become unlivable and the soul must leave. This means if somebody dreams of a neglected house with a broken door, one knows that a person is terminal ill. Dreaming of gates and doors indicate change is imminent. Walking or driving indicates slow or fast progress. Sweeping a house means getting rid of misconceptions. The activities of persons are only symbolic of the meaning it suggests—that meaning must be assumed. Dreams seldom identify individuals only their symbolic character. One could go on, but this is leading away from the issue of transcendence. But, I want to mention one other important point before going to examples. Dreams often use parables to indirectly express an idea. Parables are an archaic style of symbolic speech.

For example, a young man who opened a practice in naturopathy told me of a clear dream he had. He said, "I dreamed I was walking out of the door of my house to step into my car when I noticed that I wore a strange pair of shoes. I walked back into the house to put on my own shoes. As I stepped into the car, again I had a strange pair of shoes on. *What is going on here*, I thought, finding this game annoying, and gave up. Finally, I drove away wearing a pair of rubber boots that one can walk through mud with and can change easily."

The young man said to me, "This makes no sense. What have I in common with shoes and rubber boots?" Yet it makes sense in connection to his occupation. As a naturopath he has to understand the reason why people are sick, and literally has to put himself into their situation. Which means to walk in their shoes for a while. Actually, the dream tells him to get used to changing his shoes frequently, by suggesting he wear easily put-on rubber boots. This dream is actually advice to remain uninvolved, in order to protect his own identity. Someone is concerned about his involvement with problems that are not his own—that someone could be his own unconscious.

Because I had different people telling me dreams and had read books on the subject, I know that without openness about the circumstances of a person's life, one can not interpret dreams. Even a person's astrological sign helps to explain something of the individual's emotional background. As I have mentioned before, I frequently ask myself whether the philosophy expressed in this book is actually about the Universal Principle or is it just another philosophy. As if to answer my questions, I had a dream that seemed to confirm my hopes. Of course, it could only be wishful thinking on my part. Nevertheless, it is a dream with meaning that goes beyond personal relevance, because many other persons are involved.

In this dream I found myself and my wife driving in a tiny, agile automobile on a one-way highway, sneaking around and between stalled traffic, bypassing all sorts of expensive vehicles and some truck transports loaded with heavy equipment, that were standing and waiting. As I passed them, I knew that they deemed my effort as useless because the traffic was blocked and that one would not get far, but I wanted to drive as far as possible. Then suddenly the scene changed. We both were walking on a road in a pine forest with many other adults also walking on the road in the same direction. My wife and I walked faster than most and overtook some persons. The road soon ended at the entrance to a sprawling complex of low buildings with a wire fence around. In front of the entrance stood two uniformed security guards. As we came near, a man walking on my right side, whom we were about to pass, reached inside his coat to produce documents. Instead of stopping us, the first guard let us pass and turned to him instead. We continued to walk towards the second guard, and I thought, *What will I say, if the guard asks to see documents and credentials, as we had none?* But the guard asked for nothing and only told us to "take our shoes off" before going in. I then had only short look at the buildings which were one- or- two-stories high and well-spaced apart. It looked like a research complex or movie studio. With this observation, I awoke, before we had time to take our shoes off.

What should this clear dream mean? Since I had not had a great dream for several months, this obviously was a message with a meaning. In the dream we had passed and overtaken many better-situated persons in expensive vehicles. Were they expecting to drive there in comfort and take everything with them? As we found out, we had to walk on foot to get further, and even had to take our shoes off in the end. I took this to mean that wealth and status did not count at the end of the road—even our identity was unimportant. If taking the shoes off means to walk softly and step on no one else's feet, it must be a gentle place to go.

Was this dream symbolic of the journey of life? Since my wife and I did not continue through the gate, this was only a preview of the future. But what could taking the shoes off mean? It was obviously not a religious place, judging from the appearance.

During the following day I analyzed the dream further because I had been interested to know what awaits us after death—that could be the reason for the dream. It was showing that material possessions and status have to be left behind, which we already know, and had it confirmed. But the dream said nothing about what is beyond the gate. I gave up explaining the dream. A year later, however, there was a follow-up, which revealed more detail.

I chose to mention this dream, because it was organized and has a clear message confirming that I had left most of the academic establishment behind me. Although most dreams are personal, because other humans were involved, the message was for progress in general. Obviously, the question of whether or not I was on the right track was answered. Sometimes dreams answer questions.

Let me again return to dream symbols. Driving means advancing faster than walking. Mechanical equipment means experimental science using complex instruments and personal status. Walking, in this case, means that one has to lower oneself to physical effort. We all are still part of nature, and in the end nothing physical can be taken along. In dreams one is left without reference to a place and time, without personal identification or the ability to immediately connect the meaning to events.

One could assume that dreams are a preparation for a coming event, but without details, are meant to keep people unable to prevent what must happen. Perhaps this is what one must understand—that one should not interfere, that everyone has to live life as it unfolds and act by their own judgement of value. At least dreams are preparation for something unexpected to happen and may emotionally prepare us, which can be important to our sanity. Several months had passed when I had the other vivid dream about the meaning of taking my shoes off. The second dream was meant to clarify the first. In the dream I was at work again, with measuring equipment around me, and it was near time to go home. I said to my colleague, "I have enough today, let us close up and drive home." Then the scene changed completely. I found myself sitting in a theater watching a play, improvised by local people. When the show was over, I remembered that I said we want to drive home together, and walked out of the theater. To my surprise I had somehow taken my shoes off and was walking in my socks. So I went back into the hall to find my shoes. There were many shoes and sandals, even a few jackets left behind by others under the chairs. I could not find my shoes among them, but noticed that my shoes were back on my feet again. At once I was again with my colleague, who was waiting for me to go out off the plant. I then awoke.

This dream was as clear as the previous one, and it was obviously connected to the first, to clarify a misunderstanding that I had about "taking shoes off." I thought it meant to walk softly among gentle people and was uncertain what happened after one leaves the shoes behind. Then I remembered that during the theater performance other persons also took their shoes off. This means that one assumes another identity to mentally participate in the theater. It means to "walk in other peoples shoes" and the theater scene symbolically emphasized this idea. Thinking back to the first dream the complex of buildings in a compound, that I had seen at the end of the road could be a symbol for a movie studio. One must leave the role behind which one played before and that one has to learn a new role to play in the continuing theater of life.

This understanding was confirmed by additional details. While watching the play, I noticed that most of the people who attended the show were emotionally identifying with the characters in the play, and either had tears in their eyes or were smiling. During the play they had left their identity at rest and symbolically had taken off their own shoes and walked in the shoes that the actors suggested, even symbolically wearing other jackets. This means that during the performance they were transcending their own identity in thought.

The message of the second dream therefore was to clarify a misunderstanding. At the end of the road of life, one must be ready to leave their acquired identity and all personal values behind and to learn a new role in that studio. In fact, it could mean that one will become another person after one walks through that door. Since I awoke before it happened, I failed to comprehend the full meaning.

Does this dream imply reincarnation? Not necessarily a return in this world or in flesh. It seems to mean a change of identity and that could be on a higher level. Many people in Asia believe physical reincarnation occurs for many generations before there is upward mobility. In Europe, many people assume they will become spirits and may exist in a timeless mode or their spirit-soul goes to another dimension in the vastness of the universe. In the pattern that suggestibility would imply, one seeks the level where one wishes to be reborn. Perhaps convinced materialists who believe there is no future after death may also have their wish come true. I base my opinion on the "creative potential of suggestibility" which so far has confirmed itself.

It appears that the world is not designed from a human perspective and is not to make life easy to create values. Moral concern appears absent in nature where opposites are used to create values, where we humans are used like puppets, and where physical values and mechanical contraptions can also confuse us. How can one believe in a meaning and purpose of such a world? One must see the intent of the process of self-perpetuation. This world has evolved from a primitive past in a difficult and long process to which all have indirectly contributed. There can be no question of good or bad—we all benefit from the process which eventually gave birth to thinking man. Any existence is always more then nothing. As long as we have a chance to play another role next time, in a future time as individuals, is an experience that has meaning.

Despite a general insensitivity to suffering in nature, there is some indication of compassion with the sufferings of individual humans. This is implied in occasional dreams as preparation for impending changes to soften emotional shock. This concern implies that compassion exists at the source where dreams come from, and also requires that the beings who engage us in dreams have advanced knowledge of approaching dramatic and tragic events. This means on the next level, where the source of suggestions must originate, transcendence of time and space must be possible and is normal. Could this mean that among us some individuals exist that are partially transcendental? Are the sort of forerunners of an expansion of consciousness in humanity? Are they the persons who can occasionally see the future or feel emotions in other things, unusual persons that can perform psychic surgery, by passing through organic substances which transcends physical separation and others who can negatively influence material objects in moments of emotional stress? All these acts are only possible in a suggestible universe—a wave phenomenon that can be transcended.

We can feel consoled, however. Something out there is concerned about the suffering that the system can cause to individuals and understands human feelings, but cannot interfere or even give accurate advise of what must happen. It is almost as if things must take their own course, pointing out that real life experience is necessary to evolve to another level and that each is responsible for their part in the process.

This also implies we are not meant to be spared the experiences of pain and suffering; it may be necessary for personal development. If one can understand such reasoning, one stands above the play and with intelligent foresight one can avoid at least some emotional suffering. But the suggestion that pain and death are always present, still hangs over life and keeps us in compliance with our assigned role in the theater of living in our own world.

Recalling the first of the two related dreams. When we drove past expensive automobiles, that suggested that persons attached to material values stayed with their possessions. The persons who abandoned material values walked further on the road that leads to change. Perhaps the symbolic movie studio means that one will be given a new role to play as an actor in a new life experience. But the revelation that in essence we are actors in a play, where in the end nobody really dies should make us take life as less tragic.

The idea of immortality and advice that one should not overly worry about the casualties and suffering in a senseless war has been expressed in writing in the *Bagavad Gita* by ancient seers in India. It is implied that all is illusion, that the world is maya. Dreams are not the only place where the survival of consciousness is implied. Through modern medical technology, a number of clinically dead persons were revived and have reported that at the moment of death their entire life was shown to them in a video-like playback—from childhood to the end, as if they were meant to learn something from the experience. Many of the clinically dead persons have reported to have been speeded through a tunnel where they saw a brilliant light. They were sent back because they were returned to life by medical intervention.

The world is a value-creating process and our existence is more than just illusion. It is physically solid and must be taken seriously, but one need not be paralyzed by fear. People who have been revived said that rather than guilt, they felt unconditional love which is what suggestion of value feels like—one is valued. One has played a part in the theater of the value-creating process—whether good or bad. At that level it is the same because values are suggested by their opposites. It makes perfect sense that one should be forgiven, because indirectly one has been suggested to play a certain role and therefore when one comes off stage, one is given appreciation.

The question of good and evil comes up many times and should be addressed. In Asian religions the gods are playing with people at their will; Shiva is creator and destroyer at the same time. And the pagans of Europe believed their gods to be unpredictable. And Heraclitus, dispensing with capricious gods, thinks of nature as the "play of the aeon," as infinite change. He also said, "We are and yet we are not." To this I can only add that we are integral parts of the value-creating process, actors and spectators, experiencing interesting lives. But the individual is not the reason for the universe, the process as whole is.

Precognition

Let us continue with the subject of transcendence. Precognitive dreams happen to all sorts of people, regardless of culture or religion. Most of my dreams that are specific to my philosophical thinking are certainly inspired by interest in the subject of transcendence, but they are not a product of my own fantasy. On my own I could not have found the connections to other dimensions. Dreams have inspired thoughts of humans before I ever existed. That dreams of different persons reveal a similar pattern should be considered a form of verification that the human consciousness as bodiless spirit transcends physical limits.

The interphase with other levels is a difficult subject, but it cannot be denied. In abnormal psychology there are many cases of divided consciousness, including sleep-walking and cases of multiple personality. The so-called idiot-savants are an example because they are only partially on the physical level and often gifted with extrasensory perception. Some can solve long calculations without knowing how. That one cannot control abnormal gifts is part of the pattern of uncertainty and because is they are from a timeless mode, these persons behave like idiots in other fields. They are savants and idiots at the same time.

The concept of time is a key to unlocking many secrets. If we could know the future, then the experience of life would be uninteresting. We want the future to remain unknown and for existence to continue infinitely. This is not just wishful thinking—it is the essence of being. That is why I cannot agree with the scientific model of a meaningless universe. It insists on finality with no future at the end of life. It is not because I am a romantic dreamer, but because I am a realist, that I see a greater reality in a multi-level, pluralist universe, created by innate suggestibility to value as continuum infinitum.

The Hindu-Buddhist model of reality, in many observed facts, came to a similar phenomenal reality by implying life is basically sorrow and the world filled with meaningless striving. This negative idea, which one can understand, is experienced in the over-populated regions of east Asia. But this experience is not shared in other regions of our planet.

I know that a few modern Hindu scholars disagree that this is a negative trend, but the Upanishads and the Yoga Sutras say the goal of the wise man is to silence the mind from thinking. This reminds of "autism" as it exists in the lower levels of nature—a step backward in evolution.

Let us get back to some practical examples on the subject of dreams. For instance, a middle-aged male wakes from a clear dream during the night, in which he experiences strange pains in his abdomen. In the dream, he is imagining that he is ready to give birth to a child and awoke with real abdominal pains which is an indication that a birth is imminent. But in the dream he also knew he was a male, which did not seem to matter. He realized that this could be how a pregnant woman might feel to give birth. But after awakening, he reasoned the abdominal pain could have been caused by something he ate in the evening, which the dream conjured into the imagery of labor pains in a woman. This is a typical rationalization by normal skepticism because we do not trust our imagination. The man would have ignored it as just another dream, but during the course of the day, his son telephoned, saying he had to take his seven-month-pregnant wife to the hospital with severe labor pains. This confirmed that the dream was not the result of indigestion, but transference of an emotional experience, and was meant to inform him, a grandfather, of an important event, a birth.

In the records of psychology dreams of birth and death are common, and pain transference is typical. Dreams of the birth of grandchildren are experiences which millions of people have. So are also dreams about persons who mean something to us, when dying. Science does not deny such dreams can happen, only declares that this is outside its field—obviously admitting that science is limited to physical knowledge.

People often speak of coincidences to avoid having to admit to a design. But how many coincidences have to repeat themselves, and of the same subject, before admitting a regular pattern with a purpose? Why not make a daring deduction—when asleep, the subconscious mind is open to receive messages from other minds through telepathic thought transference. This implies that on subconscious levels we are all connected, but we only make use of it in serious matters. Otherwise we maintain individual separations and unconsciously prefer to remain autonomous individuals.

Seeking transcendence in ordinary matters is counterproductive to the separation of values and meaning. One should leave the so-called occult and spiritualism alone because it is dangerous to confuse levels. In dreams, death and birth have the same meaning. On our level there is a difference between death and birth. Dreams dramatize situations and the individuals in them are only symbolic. This means dreams cannot be directly applied to our level. A dream of killing means change. So to assume that a specific person did the killing or was killed is a misunderstanding of symbols. Actually, one knows little until an event confirms what the dream meant.

If transcendence should be left alone, why do I bother to understand and explain it? Because we have reached a point in technological progress when it becomes essential to understand reality before we commit grave damage to nature and the biosphere on our planet. Indeed, I think that a guiding influence exits, but it only interferes when serious issues are warranted, otherwise we are left alone to direct our lives.

Frustration is a powerful emotion, and has been connected to bizarre events which have been under investigation by parapsychology. There are cases of "spontaneous combustion" that have been recorded. For example, in Britain a lonely person burned out from inside, with only the limbs left. And there are cases in which a person vanished without a trace. The psychological explanation is deep frustration with life. The scientific explanation is probably the combustion was caused by an accident resulting from the disfunction of internal chemistry or an electrical discharge—not psychokinesis from frustration. But it could be that in heightened emotional excitement on a the subconscious level electrochemical separations are transcended and disintegration occurs rapidly. Without a solid shield of identity, nothing is holding the organization together.

However, in other cases of identity crisis, when confusion and fear paralyze reason, and hallucinations occur, occasionally meaningful suggestions are mixed in the content of hallucinations, indicating precognition of future events. Usually the individuals are too shaken and afraid to understand that in such emotional conditions transcendence can occur.

Transcendence has been explained in different ways by the magicians and sorcerers of old and today is understood by parapsychology as psychokinesis. For example, ancient Taoist wisdom is based on transcendence expressed in mystical style, as a few relevant sentences from the *Tao te Ching* indicate. An example is, "The sage is aloft of human affairs, and as perfect being has no self, and therefore has no name. He is absolutely happy by himself, because he transcends the distinction between the self and the world, the me and the other. He is one with the Tao and by doing nothing, all gets done by itself. This is natures way, and the Tao is the center of change." The same general idea of transcendence is expressed in Rajah Yoga, as union with the Atma, which means the worldsoul.

The statement, "the sage has no self" shows the typical requirements for transcendence, including the suggestion of doing nothing, and that all gets done by itself. This is also true when a person suffers an identity crisis, he experiences the loss of separation from others as a fear-inspiring experience. We are not meant to adopt this strange reality. To "do nothing" is to ignore the physical experience. We are expected to participate and to do something for ourselves or to starve to death. Starvation would happen if we lived in nature, but because we live in a human society, others will support the do-nothings. As self-conscious beings we are meant to participate in the affairs of the world and are given freedom and intelligence. But intelligence comes with a price. We become responsible to nature and should "act" with fairness to all beings.

Obviously to do nothing about the present destruction of nature by chemistry makes us irresponsible. Also, letting the population explosion remain out of control allows it to become a time bomb. Therefore, I disagree with religion that God will take care of our problems. We are not children any more but mature individuals. Besides, the world is not actually a place of suffering—it should be seen in positive light as a place of interesting experiences. Pain is mainly intended as warning to keep us from further injuring ourselves.

Many of the ideas and values that have come from the past cannot be taken as wisdom today. Unlimited expansion has to end sometime because our planet is already unable to sustain its ability to clean itself. There must be other alternatives to evolve than physical. Even if we tried to develop transcendence, it means that we would first have to be selfless, as it was stated: "The sage has no self." Therefore, the sage passes "freely" through the separations in the world. In our present evaluation, transcendence must be considered limited by the many levels protecting the individual from unwanted suggestions. There are only a few existing psychics and they are restricted to a particular field. This is where knowledge of transcendence presently stands and has not changed much. Various experiences by the military to develop transcendence have been foiled by nature.

For instance, involuntary "out-of-body-travel" was recently rediscovered in the West, independent from the before-mentioned Chinese writings. Nevertheless, it did not allow humans to spy on others humans intentions because it would render secrecy and privacy meaningless. Transcendence is apparently selective. It is meant to connect us to other dimensions only in times of impending crisis, as if to prepare us for a possible emotional shock out of compassion for our feelings. This compassion should not be overused as a pretext to exploit transcendence. If transcendence ever became a practical skill, the separation of values through evolution (as has been done for generations), would have been in vain, the value-creating process would become meaningless, and the universe might as well return to the void from whence it pulled itself up by its own bootstraps. That is why one should leave the occult to chose when it wants to advise us.

Transcendence can also become a burden to see and feel things that are painful as clairvoyants claim. If information is given from the other side, that is their decision, not ours. Also, "near death" experiences, resulting from medical technology, give us an indication that human interference in the process of dying is a forced interface between life and death. From this we now can assume that consciousness can exist without a physical structure.

Let us return to practical psychology. According to Freud, dreams are the "royal road" to the unconscious. Freud and all the other psychiatrists were medical doctors first, concerned with restoring health, they were not philosophers of transcendental idealism. Everyone can speculate about the world, but disconnected fragments are not a complete system of philosophy to explain reality. To achieve that, ideas have to be organized to fit into a system that is compatible with the real world. Apparently psychologists were impressed by scientific achievements and applied for recognition as a science, but failed to see that they were closer than ever to an emotional interpretation of reality.

If dreams are individual reflections of the human psyche, then the psyche of individuals is a reflection of the spirit of the greater psyche of the universe. In general, the great majority of dreams are free of moral meaning and seem to accept injustice as if it is without importance to nature. But this does not mean justice is unimportant in human affairs—it is of great importance in keeping peace among people. It is on a higher dimension of reality.

When the critical mind is relaxed, the unconscious is open to suggestion. Then transcendental images from outside of one's own field of concerns, from other places and other lives can slip past the sleeping mind and receive precognitive visions about unusual events. One is surprised when the news media a few days later reports that a similar event in the real world actually has happened. Such dreams are mostly of emotionally-charged events about disasters and calamities. According to Heraclitus: "In dreams we participate in the affairs of the cosmos." But, more than in a direct telepathic link, most dreams occur before the event. There are many historical examples of well-known transcendental seers besides Michael Nostradamus and Rasputin. For instance, Edgar Cacey, a healer, thinks that departed souls travel far beyond their present location and that after departing the body, the soul first rests for several decades or centuries. The soul remains in a limbo before being deployed again in the physical dimension somewhere in the stream of time. From other accounts, one may conclude that one is attracted to a life-style one wishes to experience and thereby manifests oneself again where one likes it.

Perhaps the general idea about the meaning of being was never intended to remain a great secret. Many persons with common sense think that life is meant to be an experience. Poets such as Shakespeare, Goethe and Ibsen wrote plays about the meaning of life. In fact, we have turned into a spectator generation, wanting to experience more than just our own life by sharing the experiences and dreams of other interesting lives. Indeed, dreaming is similar to going to a theater. While sleeping we participate in the affairs of many other's lives—that is what dreaming is. One will find out that dreams are not often meant to reveal the future before it happens, but may simply be entertainment. The future is still being created in the continuing value-creating process that is flexible to changing conditions.

The ability to adjust to changing conditions demands self-orientation to roughly sense what is required. The orientation that is in everyone has that elusive quality which may best be called "innate suggestibility to value." It must already exist at the level of waves before the particles combine to organize the physical manifest reality, which we perceive as the tangible world. This is the world in which we live and never needed to know how it functions or what caused it—unless one is curious, which the reader obviously is.

Evolution expresses itself in multiple ways through the organization of ideas. We are unconsciously part of the world of ideas. We think in ideas, about which dreams can reveal more. But whether one has achieved the goals to which one aspires is partly through chance. Playing the game is the main issue—the goals are always a step ahead, still to be reached—and it will mark a new beginning.

To understand dreams requires knowledge of some of the values and individual circumstances of the dreamer in order to interpret the dream. A woman told me a true story that was preceded by a dream. She said, "I dreamt that I was in a kitchen which was in need of housekeeping. The dishes were piled up in the sink and there was disorder everywhere. Then the scene shifted to the outside. I was given the general overview and could see a modest house over which a small noisy airplane flew. Out of this house a woman came running with her head on fire. I awoke with a clear recollection of this dramatic scene thinking that something must be in progress. I just did not know what." I had a dream book on hand that said that a house means "the person" running out of a house then should mean "being out of oneself." The head-on-fire clearly means a mental breakdown. The scene with the airplane adds noise and location. Sure enough, three days later, the woman said, "I was asked to take over the household of a mother of four young children who was taken to a mental hospital. She was actually a relative-in-law of mine who came from a farm to live in a city. She had lost her husband through his accidental inhaling of poisonous chemicals while spraying his fields." The farm had been sold and the young mother moved into the vicinity where the lady lived who had helped her to relocate. Outwardly, the country woman seemed to manage by herself in the city environment. "Since I also had three children of my own and a husband to take care, I was involved in my own household and expected nothing serious, until the dream preparing me for the breakdown. That the purpose of the dream was meant to prepare me for a shock is a logical deduction. I now had to take care of two households with two sets of children—her four and my three. I managed for a week, driving back and forth, until a homemaker was found to help to take care of the home and the confused children during the daytime. The youngest child was only one year of age and the oldest was seven years old," she said.

This is an example of a precognitive dream, preparing the dreamer for something unexpected. Just for the record, the young mother was cured with drugs, was released, and also received some psychotherapy. Eventually, she remarried and found hope again and could see a future for her family. One important point should also be mentioned. Simultaneous to this dream, a letter was received from the country woman, saying that she had a dream, in which she saw the end of the world: the sun turned black and darkness came over the world and the stars fell off the sky. The dream meant that one should pray and repent and prepare ourselves, because judgement day is near. This is typical to persons with religious values who are experiencing a mental breakdown. They often mistake the experience of the sky falling down as the whole world coming to an end when it only means that their personal world is disintegrating. Because of this common dream I am skeptical of all prophets wanting to tell us that the world will come to an end. It could just be that their negative expectation misleads them to assume the world is like them.

In this case, it was the hopeless perception of her personal future that overwhelmed the mother of four. But the second cause was the self-suggested guilt as she apparently blamed herself for a part of the events that took place. Among modern theologians, the emphasis is away from sin and hellfire and towards love and forgiveness. But fundamentalist religions still preach eternal damnation. People who live in remote areas still believe in old-time religion and sin, suggesting guilt.

Jung was correct—many cases of mental illness have religious origin, implying that guilt is suggested. Yet, who would not have a cause to feel guilty in the eyes of religion? That is therefore only a part of the illness. We all are made sinners collectively at birth, so we became responsible for the sin of Adam and Eve—yet this could not affect everyone. The reason was a loss of hope. Her future fell apart and caused the mental breakdown.

Let us change the subject. My interpretation of dreams is strictly philosophical—to understand the "psyche of nature," not to cure emotional problems. Since I lean toward Jung's individual psychology, several symbols are pointing to transcendence. The symbol of a house or building in which the individual soul lives implies that the spirit considers the body as a "temporary dwelling" only and can depart from it when it becomes worn. This suggests that the spirit-soul transcends physical limits. In dreams we switch scenes, and transformations happen frequently, as the dreamer experiences disconnected bits of scenery that make little sense. At further inspection, this is the symbolism of the picture language which is very relevant to understanding the psyche of nature.

It is amazing how far and wide individual dreams may occasionally wander. Normally dreams relate to the interests and values of the individual. This means dream symbols are individual in detail, but many are still common to all. The dream symbols repeat themselves in all human dreams, and so do the emotions which go together with the symbols.

Another common dream symbol is "water," which means life force. It is also the basic substance in the cells of all living things. The "door" is also symbolic of change, and fire of emotions, birth and death are similar in meaning, as if they were connected in a circle. The language of dreams also uses parables or puns which contain an indirect message, leaving it to the intelligence of the individual to make sense of its meaning. Consequently, the perspective that I take of dreams is based in part on the information of the individual's life and the emotions that are experienced.

For example, here is the dream a stranger once told me: "I am in my house, sitting in the living room and a flood had occurred. The water is rising, the furniture is floating around, chairs, books and all. I feel helpless. With this memory I woke." I knew then, from his experience, there was a crisis involved. He admitted having disagreements with his wife. This explains the flood, that symbolically threatens his house—his person. The furniture and books mean his values.

The house as symbol of a person, and water as a symbol of life were present, suggesting flexibility to the dreamer and a compromise. The dream also suggests his values may need adjustment, and a compromise with his wife is indicated, to prevent another dream of a crisis. There is one consolation—floods are only temporary, the water will dissipate again and the crisis will pass. Healthy marriages are kept interesting by mutual interaction. In marital relations we observed that the husband stands for an existing thesis, and the wife becomes his antithesis. Together both evolve through a continuous chain of confrontations by reaching synthesis after synthesis. Tolerance grows and reason can evolve—if one listens to the partners concerns and makes amends.

Freud is correct that dreams reveal unconscious problems and repressed wishes, and that many dreams are wishful thinking, often meaning the reverse from what they seem to suggest. As to the dream about the flood, Freud's sexual theory does not apply nor does Adler's social interpretation nor Jung's spiritualistic explanation. I think that the flood dream shows a conflict of individual values and philosophical ideas. The dialectical principle of Hegel better explains the synthesis of psychodynamics. Philosophy has a larger perspective of reality because it includes the physical organization and material values as well.

For Jung, life had a spiritual dimension and the purpose of life was individuation. Life means to come to an understanding of the duality within oneself—to find a balance between the bright and dark side which we all have. In other words, Jung's *Analytic Individual Psychology* is related to the dynamics of the value-creating process, although the existence of innate suggestibility of value was unknown to Jung. This connects Jung to philosophy. In fact, the entire idea of the unconscious was first exposed to light by the philosopher Friedrich Nietzsche, although the development took a turn to suggestibility, away from "will" typical in German philosophy.

Going back to dream symbols for a moment. Because the symbol of death is followed by birth which are interchangeable in dreams, one could think of life as a endless circle. Nietzsche called this circle the "ring of eternal return." He referred to certain flashes of memory that one occasionally has as having "seen before," better known as *deja vu*. Nietzsche revived the ancient Babylonian idea of the great year, that suggested that there are cosmic cycles of the same world and the same life experience repeating itself. I am not denying that the experience of deja vu occurs, but think it can be explained differently, namely by remembering scenes out of dreams. It is unlikely that the entire cosmic process should have to repeat itself in every minute detail, over and over again like a mechanical clock. Instead, one remembers a scene from a recent dream, in which one has transcended forward in time by a few days and therefore is familiar with it.

In fact, I have had several such deja vu experiences myself and like to mention one such dream. I dreamt that I was in A large room in which ancient Egyptian mummies and artifacts were displayed. I noticed that the ceiling had a skylight, and to the side adjoining this room was a large stone relief on a wall, depicting a sequence of events of ancient history. I was attracted to this relief, but as I walked over I awoke, remembering the details of the dream clearly and even wrote it down. Weeks later on a flight to Europe, I stopped over in London for a few days to see the sights. Among others, I went to the British Museum, where I had a deja vu experience. There was the already-seen room with Egyptian mummies and the skylight in the ceiling, yet I remembered that I had seen this room in the dream, not in a previous life. I walked over to look at the stone reliefs on the wall, to find out what my attraction was to the relief. The stone relief was from ancient Assyria, from a time when the life of most humans had far less value then it has now. Inscribed in the stone reliefs were the heroic deeds glorifying King Ashurbanipal. The relief was excavated from his palace at Nineveh. On it were scenes depicting the slaying of lions, of enemies in battle and also the beheading of prisoners of war, after they had surrendered. Male prisoners were commonly slain because women and children made better slaves, and men could not be trusted. Such memories of a cruel and inhuman past should remind us that we have improved since four thousand years ago. Soldierss who give themselves up are now spared and slavery is outlawed.

It seemed that the stone relief had some advice to give, namely that we are progressing, and that soldiers that surrender are not executed. However, a few such inhuman acts still occurred in present times. In the latest Balkan War, in 1995, prisoners and male civilians were executed as a form of genocide. But officially mankind has accepted humane treatment of persons with peaceful intentions. Although concentration camps are a dark side, as is the ethnic genocide of innocent individuals, they are in the past and one must forgive because life has to go on.

I am not surprised that many people think that "God is dead." The reason for brutality in the world could be its opposite—to acquire compassion for all beings who share our planet. Fear and frustration are ways to prepare human beings for transcendence and awareness of other dimensions. To be open to transcendence also has a negative side. It is not good for our sanity to be without protection. Suggestions from other entities can influence and disturb our lives. One has to remain stoic and keep a separate identity in order to stand life. To know that life is only a temporal experience should help to us understand that existence is not in vain. We are intelligent beings who can avoid suffering to some degree and that makes life tolerable, if not an interesting experience, which at such moments seems difficult to accept.

Informative dreams are only meaningful to reasonable individuals with open minds. This indicates that intelligence is an advantage, and nature rewards it. Of course, in all walks of life intelligence is its own reward. When mankind chose to develop reason it made a wiser choice than did animals who chose to grow bigger teeth and to have greater strength.

One must conclude, that nature is not meant to resolve good and bad or bring about justice to the world. Nature has only one single purpose—to create interesting experiences where otherwise nothing would be. Morality may not be the main issue in the world process, but it is a major issue in human society. It was through organization that mankind became dominant, and organization depends on willing cooperation. Justice is a necessity for any civilization to function. Human intelligence wants peaceful cooperation, which inspires morality and reason. We are what we make of life. Life is imperfect but better than nothing.

In the first chapter I mentioned fire-walking and psychic surgery as examples of transcendence which involved passing through fire as if it were unreal. These events require one to go to physics for an explanation. After electricity was discovered, physicists began to think in waves lengths and frequencies. While calculating nuclear energy Planck discovered that a certain pattern repeated itself which occurs in multiples of a constant quantity of energy. Planck called this entity the quantum, the ultimate indivisible energy particle assumed, which only has numerical value to physics. On the existence of this invisible entity the quantum theory was based. It was a mathematical explanation of wave frequencies organized in bundles also called particles or wavehicles in nuclear physics. In particle physics everything is quantified in frequencies and wavelength and spine of pure energy. One should think if wavebundles come in multiples of a constant entity that means they are organized, which implies they are conscious and in large fields are emotionally excitable. The organization of particles becomes physical and we note that atoms have qualities besides numerical energy value. For mathematics, the qualities are secondary. For living beings, the qualities are of fundamental importance. Qualities have meaning for our existence because they are values for us. Waves manifest as they organize as atoms, such as hydrogen and oxygen. And they organize as molecules of water which has direct value to organic life. Other atoms also have qualities used by nature to organize into all the many things and beings in the world. In short, this electrical wave phenomenon has organized itself in a value-creating process with ideas of individual identity. Yet, the original wave nature is there. The transformation has occurred as synchronization to meaningful values. This means reality is a phenomenon of synchronization by high-speed vibration of waves. In particular, there are cases in which the speeds of the vibration between two objects becomes equal and stand still next to each other—these are windows for transcendence. This is the general idea of synchronization and the matching of values are complex in detail. In other words, transcendence means that one could, at certain emotional moments of excitement, pass though solid objects. In fact, interpenetrable waves that occupy the same space, but are only perceived if they have value for a species or mode of being. This would mean that things are perceived on relevant levels such as the physical world in our range of perception. In theory, we understand how transcendence works, and that the vibrations are emotional excitement. This explains how suggestion can organize conscious waves in which we experience meaning as interesting worlds.

The fact is that suggestion of value has been proven to organize minor cell formation and during high emotions, excitement can influence material objects. We are also aware transcendence is limited and for a good reasons. We cannot be trusted to act responsible. As emotional humans we can be discontented with our existence and then become psychotic. Nature is resolute and replaces misfits with willing participants.

The exact details of transcendence are unknown. Psychic surgery apparently is explainable as positive emotional influence on the organization of the body of a patient, which allows synchronization to occur. In principle, it is suggestion that influences organization on the physical, as well as on the mental level and allows a restricted transcendence of the barriers of body and mind.

Let me offer an example of the transcendence of time on the mental level. The following dream is from a woman who usually is skeptical about the occult. After she told me about a normal dream of precognition before her mother died, in which she had a deja vu experience, the question of linear time came under discussion. I explained to her that time belongs to the separations of values and meaning, and physical separations can be transcended in death, but mental identity seems preserved. I asked her whether she had received signs from her departed mother, existing somewhere in another dimension. She said: "Yes, I had several dreams in which my mother was present. One of them in particular impressed me. In a dream our entire family was standing up to our hips in the flowing water of a river. My late father and late mother were there, also all three children and my husband who are still alive. I had a conversation with my mother, and remember asking her about the 'meaning of life.' She explained it to me. It was so simple and made good sense. After I awoke, I remembered the dream in detail, except the answer about the meaning of life. Only this information was wiped from my memory. Probably my subconscious knows it and my conscious does not want to remember and prefers uncertainty."

The symbolism of this dream, standing in the river of time, is confirming the impression that we are trying to establish a temporal identity against the stream of change, to create meaning where otherwise nothingness is the ultimate reality. Transcendence has a necessary function in the design of reality. It provides the internal connections to hold all the separate beings and entities unconsciously together. Transcendence is a more than mental gravity, it is the loyalty the holds the stars in place and all depend on it so that tomorrow will come.

This family meeting in her dream points to inner connections in families as loyalty beyond death. I assume that the reader by now might have a general understanding of the Universal Principle. It could all be said in the simplest of words: All there is, is innate suggestibility of value, making waves in nothingness and thereby creating the phenomenal world. As humans we are somewhere between levels and have to make the best of it. To be imperfect seems still better then nothing at all. Besides, the process of transforming quantity into quality provides hope for further improvements of human existence in future lives.

11

Consensus and Scenario

IT HAD TO HAPPEN EVENTUALLY, and it did happen in our century, that the classical mechanics of Newton were challenged by facts—first by Einstein, but more seriously by quantum theory. When it proved impossible to combine gravity with quantum theory, there was a problem.

During a vacation, Werner Heisenberg found the solution, namely that mechanical and kinetic relations are incompatible with quantum fields. He concluded that patterns arrange themselves as matrixes to synchronize with wavefields. From these discoveries I deducted that matrixes could be organized by the same innate suggestibility which I knew organized cells in living beings and synchronizes meaningful ideas.

All this came together as a theory of the universe, that innate suggestibility is already present in primal energy and reacts to observation. This was also observed by Heisenberg. I concluded that the existence of galaxies could be the result of innate suggestibility to organize fields which emotionally excite themselves into a frenzy of wave motions, as the light from which the creation of worlds through ideas has resulted. Organization by differentiation into meaningful things and beings also involved chance as evolution by trial and error shows. This implies that elementary consciousness evolves by organization of value. Quanta fields may rest in apathy or can be become emotionally excited and congregate into galaxies to create values. As mass-emotion they become the light, bringing illumination to the darkness of the void.

In the primal state of consciousness, these highly suggestible and excitable entities excite each other into emotional frenzy and ignite into fiery chain reactions that create waves and light. Waves organize into subatomic particles, which organize into pulsing atoms that we perceive as material. It may be a state of exited happiness in which these entities become inspired to organizes worlds because suggestion of value is a positive inspiration. They certainly generate the energy to form multiple stars and planets on which self-conscious beings have evolved by organizing themselves into meaningful designs. In particular, the human species is organized from conscious cells which are suggestible. In a hypnotic trance, humans can be influenced, but only to a restricted degree, compatible with the unconscious identity of a self. Otherwise, suggestions are expelled. This restriction leaves the unconscious identity protected against outside manipulation. In fact, all of nature is protected by countless barriers and separations against unwanted suggestion. This implies that a system created by suggestion that protects itself against outside suggestion from interference on lower levels can appear unchangeable. But the general process of evolution indicates upward mobility to expand consciousness to increase the experience of meaning, particularly on the human level.

Consciousness must also posses orientation and reason to recognize opportunities that suggest meaningful directions to evolve, namely to sense possible value. Judgement of value is the same as the word reason. Philosophy has long ago recognized that values have evolved from opposites. This means that values give meaning to each other, thereby creating each other. Opposites such as up and down, good and evil have meaning, but only as pairs. This has let me to think the universe is a value-creating process. I see innate suggestibility to value as the single cause for the necessary dynamics to keep the universe in motion—emotionally moved that is, and indirectly influenced by the conscious entities that desire it.

If there is one single principle that cannot be contradicted, it is that meaningful organization occurs by innate suggestibility of value and meaning, and that organization comes to self-understanding by insight. All else is unnatural and misleading. Conflict, however, may have a purpose in the dynamic to make the world interesting but meaning is the cause.

Because evolution is highly organized, nature requires the presence of basic consciousness. The reality of organization in nature disproves accidental evolution. Experiments staged in the laboratory have no resemblance to natural processes; the mechanical model of the universe is unreal. At least it is electronically expressed.

Philosophy and theology never believed that the primal energy should be mechanical and accidental. However, because scientists could do things by forcing chemical to synthesize they assumed this is how the universe was born. Philosophy was left to explain nature by metaphysics, and theology, as the creation by an unpredictable God. Science and technology gained a dominant position because they used force and produced instant results, whereas nature took much longer to evolve by internal organization. Recently, a large segment of thinking humanity became aware that it is not by force how nature creates, but by slow organization. It should become clear that consciousness is innate to primal energy and that organizing ideas are creating new values. Besides, all physical structures are temporal and finite. Only innate suggestibility as consciousness is eternal and infinite and it is continuously reorganizing itself to produce interesting experiences.

This explains that the universe is infinite, but temporal details are finite because they are being constantly renewed so that we can also experience interesting lives and changing events. In short, the universe repeats finite experiences within the infinite existence of consciousness. This also means all wave manifestations, including the human body (a finite structure) are part of temporal wave synchronization. From a general perspective, the many finite worlds are the infrastructure for consciousness to experience a physical identity. Therefore, the purpose of the value-creating process is to experience meaning. We know that life is a physical commitment to participate in what is the "world theater," with many shows running at the same time.

This understanding came together from many sources, yet only in our present time has it become confirmed by evidence which culminates in unifying a massive number of details into a single principle. Before I became aware that during hypnotic trance cells follow orders from outside, I knew only general bits of information about extrasensory phenomena and electromagnetic energy. When I found out that suggestibility influences organization, I made the connection that emotionally we have evolved from the same common source which connects everything—namely innate suggestibility.

But we humans have only minimal influence on organization by suggestion because evolved identities are protected from interference by nature, which reserves the right of the creation of values for itself, probably through differentiation by evolution over millennia. This implies that all beings are separated because their values are distinct. They become estranged by incompatible values and their differences become dramatized, even though on the emotional level we are still connected. During sleep, when we lower our critical faculty, we open ourselves to other human experiences through dreams. Most people do not remember their dreams—occasionally parts are recalled but they are soon forgotten. Only very rarely do entire dreams come into full consciousness.

One can call emotions positive or negative. On the human level, emotions make a person to feel self-confident or meek and worthless. Positive and negative are emotional states, typical of wave energy which affects the entire value-creating process from organization to disorganization of matter and mind. This explains why transcendence is possible if one is open-minded and feels self-assured of one's value. But, it is not just a question of saying that the mentally ill are out of synchronization. When they are in a negative state, they have closed themselves off to synchronization.

As we know, without drugs, intolerant and critical persons would never achieve transcendence—for this genuine appreciation and tolerance of different values is a prerequisite. This explains why psychic healers and psychic surgeons are gentle persons. Doors open to them which stay closed to intolerant persons. This explains why critical, objective studies of physical things reveal little about the existence of innate suggestibility, because countless separations protect the structure of the world from intolerant intrusion.

The evolution of innate suggestibility in organic structures is by itself a parallel to the evolution of matter. The difference between mind and matter is only in the degree of consciousness. Materials are withdrawn on a passive and autistic level, giving humans the perception that materials are dead. This belief allowed materialism to arise to what it is. Yet, this view is already obsolete because physical substances are also emotionally unpredictable.

Many philosophers, such as Schelling and Hegel have concluded that the human species has reached the level of self-consciousness to live by reason. Two hundred years later I took the next step, to explain that the purpose of the universe is not just to develop rational humans in which consciousness becomes conscious of itself. The greater purpose of nature is a general value-creating process by differentiation, to experience meaning where otherwise nothing physical is—nothing physical but consciousness which is immaterial.

In principle, this explanation is based on a psychological concept of reality, which corresponds with the emotional psyche of individual man. In detail, many versions of values have evolved, which have confused the specialists. In particular, the limits of suggestibility have confused the philosophers, to consider suggestibility a mere side-effect. Obviously, a system based on suggestion must protect itself against unfitting interference from negative suggestions by installing countless barriers and individual compartments to ensure the separation of values and identity. This resulted in the physically divided structure of the differentiated universe. We know this exists physically, and we also know that individualized consciousness exists because we cannot deny it without denying ourselves.

From this new interpretation, an emotional cosmology revealed itself. It fits the existing facts in the real world, on which a new model of the universe should be based. The existing facts confirm that the dynamics of the world is inspired by the emotional and the physical interaction of all entities that together created a highly complex nature which in detail is random and individual, but in general follows one single intent, to provide interesting experiences. The quantification of physical objects may have helped in the mechanization of our civilization, but mechanization is not the purpose of the world. It is rather an unnatural distraction from the evolution of consciousness. If there is a chance for man to travel through the universe, it is by transcendence of space and time. Mechanical means are unrealistic, considering the distances involved and the shortness of human life.

Besides, quantification misses quality, which is the real issue of a value-creating process. It also ignores the development of socialization which is important for the willing cooperation among humans. Socialization is one side of the civilizing and moralizing process of mankind. This implies avoiding confrontation, a direction of progress on which evolution is already committed, and we humans have already embarked by global cooperation. The other side is its opposite to maintain separation and its implications—loyalty to values and an established identity. In a larger sense these are the two opposite poles within everything, and they are the opposite poles of a magnet, attraction and repulsion.

We already know that we have evolved to a higher level of consciousness and insight, and now even understand why this is happening. The only way that transcendence of separation is possible is by admiration and appreciation of other beings. This opens doors that are closed to self-concerned individuals and the critical approach of science. It may seem that religion has a point, but not as long as the religions are divided, and insist on dogmas making them similar to political and commercial interests that concern the majority of mankind. Yet how can one live without concern for personal advantage? One certainly can, by only taking enough that one needs, and do it without malice and contempt—this other beings can understand. I feel that one should be open-minded and fair, and responsible for what one does.

Inadvertently, open-minded and tolerant individuals are favored by nature with understanding, while intolerant and competitive person may often acquire material riches, and seem successful. But the material level is limited and transcendence to other levels is contingent on tolerance and fairness. Obviously, the direction of evolution aims to achieve greater refinement of consciousness, emotional and morally by evolution of tolerance.

This is not a personal belief, but one deducted from reality. If one thinks this perspective has a taste of religion and morality, one is mistaken; there is no place for sin, guilt or fear of punishment in the world theater. In fact, most religions are intolerant, dogmatic and unforgiving. Nature is a spiritual concept, but not of holy spirit.

However, those that believe that reality is a wave phenomenon are in agreement with nuclear physics that the entire universe is without material substance. It is true that humans and all other beings are suggestible to value. This implies that the spiritual essence is suggestible to value. Most rational humans agree that nature is stratified and the levels have been confirmed by scientists. Most also agree that evolution is progressing by transcending and expanding these levels, and that it happens by avoiding confrontation if possible. Evolution seeks novel ideas to bypass conflict—to change to alternative modes that are more intelligent. It is by transcendence that evolution and the human mind progresses. We will continue to evolve, individually, by transcending. Science and technology discovered that the material level is limited and eventually we will reach the limits of mechanization. One can transcend time and space in dreams without going too far from one's values. The positive form of transcendence is natural for open-minded persons with a selfless outlook on life. At the top of the list are psychics and healers, with the psychic surgeons in out in front. The reason why not many such persons are well known, is because selfless individuals are rare in our materialistic civilization. Or perhaps fearing ridicule, many keep their psychic gift to themselves avoiding conflict.

This brings us back to average human experience. From recent knowledge of the psychological side of life, we have to admit that the influence by suggestions on the psyche of man is real. And one can explain the unconscious orientations to meaning in the organization in nature, if one extends suggestibility to all of reality. This is exactly what I did. I realize that many might not want to go so far out without proof. Yet how much proof is enough? Nevertheless, we exist on a physical level, and materialism has a valid point while we are on that physical level. Mechanical gravity applies.

The most recent confirmation of innate suggestibility comes from "brain research"—the hottest subject in the last decade of the twentieth century. Scientists measure electrical waves in the active human brain as it responds to emotions and thoughts—only possible because of the latest microelectronic devices. The human brain is responsive to demands and has organized itself as needed. If we use our brain, it will add neural connections and expand; if unused the brain will reduce its capacity. This same pattern of organization (by demand) applies to the entire organism, which will adjust to changing conditions. This confirms suggestibility to value. It is certain that the brain is not mechanical, but synchronized to matrixes of meaning.

As many others, I think that the ideals and goals and role models that one subconsciously adopts, influence development and personality. The question here is: Why would waves organize themselves by innate suggestibility into diverse worlds? The answer is always the same: To be something of value. To be anything at all is more interesting than nothingness. This clearly reflects what evolution tries to be, anything and everything. Ideas known by man have been expressed in all things and beings since ancient times, long before suggestibility was discovered. Socrates knew, at least in principle, that ideas are real. In the more recent past, people have organized their own little worlds by ideas that suggested value and meaning, just as we still do in the present and will do in the future.

Physicists know that reality is a grand wave phenomenon with countless factors interacting, but they believe it all is accidental, without intent or meaning. When suggestion did not work in practical experiments it was disqualified. This is why most gave up on suggestibility as a theory of the universe.

Protection from interference is necessary, especially in electronics which depends on the separation of waves to send various radio and television programs into space. These waves must be able to coexist with countless other signals without confusion. They all are interpenetrable without disturbing each other, and protected by separation of identity. That is why suggestion is limited to individual self-suggestion.

The understanding of autism, first noticed in children, is also a recent discovery in psychology. It is marked by obsessive repetition and cyclic activity resembling mechanical functions. However, emotions and basic self-orientation is still present. In fact, it is also present on the material level, reminding us that we are only a step higher in consciousness. Obviously, the design of nature by levels is organized in such a way that the lower levels remain self-absorbed so that they ignore the next higher level, and remain indifferent to exploitation. The repetitive behavior of physical entities like the stars and planets show typical autistic behavior. They are self-absorbed in their repetitive tasks. This shows suggestion of value to an assigned role, through loyalty to itself. It is an explanation of the elliptical orbits of planets around the mother star because mechanical laws fail to account for these orbits.

Philosophy may depend on science to discover the details, but philosophy is ahead of science in understanding of reality as a whole.

The fact that the present scientific model of a mechanical and electrochemical universe is in contradiction against the equally known fact of meaningful organization, proves that the model is in need of enlargement and return to nature.

Ultimately, innate suggestibility to value is the indirect cause of causes, which repeats itself in principle everywhere from invisible waves to thinking humans. The desire to be different is expressed as individuation, which causes the separation and reorganization to greater value and meaning. Because the organization of higher evolved beings is inspired by ideas that suggest value, one can understand reality as the evolution of ideas, besides being a value-creating process.

If one compares the harmony and discord of sound waves to life in general, the world could be a grand symphony of experiences in which emotions and feelings are expressed musically. One can imagine that other values in different modes contribute to what could be a symphony concert. It is created by the emotions and organized into sound in the mind of a composer, indirectly expressing the emotion of the psyche of the universe as art.

This all means that evolution of organic life is a separate mode of organization, within a larger process of many modes and realities. Indeed, there are many examples in nature of processes within processes. Evolution is a grand example of cooperation in the organization of countless diverse beings that all have a particular idea of self-worth and unique individual experience of meaning. The question arises: Are we leading nature into a dead end? It is possible because materialism is ignoring consciousness and technology is disregarding nature for short profits.

Many scientists became attracted to Asian philosophy, which is totally spiritual. Buddhism, for example, has always had a negative perception of the meaninglessness of life and the world in general. Some scientists are attracted to this concept. But I cannot accept that all human efforts are meaningless, and the brutal universe is in vain. It is a negative concept that lead to nothingness.

Allow me to disagree with this negative train of thought that caused some scientists to consider the Buddhist way of thinking. To give up on the world means abdicating our responsibility as intelligent beings. Evolution has given us self-consciousness and the discovery of nuclear energy gave us devastating power over nature, making us responsible for the future of the biosphere.

From an organizational perspective reality is a process of transforming "quantity into quality." What scientists and philosophers, what theologians and mystics each saw as different, exclusive realities, were only levels and fragments of a far-larger reality. Seen with the mind, the universe has two main trends: one is to differentiate itself, the other is to unify. In physical terms, centrifugal and centripetal forces are in a state of balance.

Human psychology tells us that if we want to be separate individuals and to feel important, whether one is also important to others is secondary. We first must be loyal to ourselves. When people primarily seek the acceptance of others, they expose themselves to the caprice of others. Many people understand the wish to be different mainly in the material sense—as egotism and greed—when it should be understood as assertion of personal value. The value to oneself is most important, and if achieved, can shut out the rejection of others who seek to destroy a person's self-worth.

Mankind has evolved to a level of reason and tolerance, despite temporary throwbacks to an animal mode. Rugged individualism that is common with materialistic values should not be mistaken with individuation to higher consciousness. Intelligence has its own reward. It can foresee meaningless pursuits in useless objectives and avoid them.

The transcendental connection of the parts to the whole was implied in ancient times. One must assume that speculation and transcendental insight was the main method of knowledge of the magicians and seers in the mystical ages.

How can a mere part understand the meaning of the whole? Because we all are part of the same consciousness and psyche, the only difference is in the structure of organization. The outward physical and chemical details to the organization have been explained in part by science, but the inner organization has remained a mystery. It is better understandable by psychology and transcendental insight.

Observations of the heavens by astronomy have revealed that galaxies sometimes devour one another, that stars occasionally go supernova in a last burst of mania, and that black holes seem like deep emotional depressions of the cosmic psyche. Such behavior is comparable to emotional reaction in humans, but in principle only and on a different scale and mode. This are only some of many indications that the universe is not a mechanical, but an emotional process.

When I became familiar with dream symbols, I was amazed how clearly the symbols fit into a pattern that could explain reality as organized from the unconscious by suggestions. The unconscious expresses itself in a mixture of symbols, numbers and words, using all forms for conveyance except direct explanation.

As I was coming near to the end of this book and had passed the mature age of seventy years, I thought there was nothing more to prove and wanted to call the book finished. Yet, I was aware that I had not explained why and when science took the wrong turn to mislead humanity to mechanization. This I found out in a clear dream.

The information came one night when I woke from a clear dream and remembered that I had asked a man for information. I was at a railway station, perhaps in Austria. I asked this gentleman about rail connections, he answered in reserved, precise German. He suggested that a "connection via Bregenz and Prague" would take me where I wanted to go. With this specific information I awoke. It took me a while to make sense of the dream. The town of Bregenz meant little to me. It was just a place in Austria on Lake Constance, but I knew more about Prague from history, since I grew up in Bohemia. I begun to ask myself what was relevant for philosophy in Prague? It could only mean the short period in the Renaissance when Prague was the intellectual center of Europe, where scientists, philosophers and alchemists enjoyed the protection of the Emperor Rudolf II. This was at a time when everywhere else the church prosecuted dissidents who openly dared to questioning the divine dogma. Gathered in Prague during the beginning of the seventeenth century, were more than two hundred intellectuals of Europe, of all persuasions and beliefs. Among them were astronomer, Tycho Brahe and the mathematician, Johannes Kepler who were enjoying the hospitality and protection of Rudolf II. Rudolph II had inherited the Bohemian crown, and the crown of Holy Roman Emperor, the protector of the Holy Roman Church. Following this lead, I remembered that at that time Kepler had calculated the orbits of the planets of the sun as elliptical which had contradicted Galileo's expectations. Of particular interest for my philosophy was the difference in thinking between Galileo and Kepler. Galileo strongly believed that Copernicus was right with the solar system. But he disregarded Kepler's findings because based on his experiments with free fall, he was convinced that nature is mechanical. Galileo assumed the stars followed mechanical laws which never varied. Galileo is rightly considered the first experimental scientist and is still correct about gravity on our planet. But the extension of gravity to the orbits of planets and stars in space was wrong. There is no gravity in space—astronauts have confirmed this fact. This hint from a stranger in a dream gave me the chance to challenge the credibility of science right from the beginning. No doubt, this saved many arguments about my credibility. The question was answered when mechanization started—namely four hundred ago with Galileo. The fact that I had received a hint from a higher level is important. It made science human, and open to error.

From the presence of unconscious organization in nature we can deduct that nature is organized by a higher intelligence. Our body certainly is not organized by our conscious mind, nor could animals or plants have the intelligence. If we listen to our unconscious we may gain some understanding of what the world is about. I think that nature is giving hints, if one seriously wants to know. This is why I think that we have a transcendental connection in dreams, and believe that we "all" are connect to a greater mind field. The unconscious has a general understanding of the purpose of life—that is why extrasensory insight helps us when logic fails.

The unconscious connection, as transcendence, belongs in a system of reality that goes beyond the material structures. It is part of a grand wave phenomenon, of which materialization is but a means to suggest realism and to force commitment to participate in the experience of meaning. From my experience with materials, I presumed that materials have a basic self-consciousness, and the assembling of machines from many parts does not create a new identity. This a rather rapid process when compared to gradual organization in nature, in which consciousness has had enough time to assume a united identity. In fact, for a human being to change identity takes time. For example, it takes the first five years of a child's life to become conscious of who one is supposed to be.

Without unconscious connections to a deeper source, we would never grasp abstract ideas and understand our assigned roles or comprehend each other's thoughts and intentions without underlying thought transfer. Subconsciously, we often understand each other too well, but cannot interfere in the lives of other persons. If we did, it would not be their life experiences. This is why beings on a higher dimension refrain from giving direct advice—everyone is meant to experience their own lives. Individual independence is the whole issue behind separation. Only at moments when one becomes uncertain how to proceed does one seem to receive indirect advice. That's when transcendental dreams can help.

One such dream, mentioned in the chapter on evolution, happened when I developed moral doubts about the waste of life. In that dream I wanted to add a dormer to the model which I had completed. It was rejected as superfluous addition, implying that evolution evolves towards tolerance which is the basis for morals. The second time this happened to me in a dream, was when we were driving in a small car past the materialistic majority. And in order to progress one had to leave everything behind, and had to walk on foot to a gate at the end of the road. In this dream I theorized that the request to take the shoes off, meant behind the gate was a gentler life on a morally refined level, like a heaven. This again was made clear to me by a more explicit sequel to this dream—in which I was taken to a theater and shown that to take the shoes off meant only to start a new life experiences and to leave the past behind. Moral judgement was not even implied; it is not an issue.

If this idea is applied to the world stage, to the political theater, then it would appear that certain persons were assigned important roles about which they may not have had much choice. Sometimes there are catalysts introduced into history to cause change. One such catalyst for the advancement of knowledge was the controversial figure of Rudolf II. There were many others in human history like Constantine the Great who supported the idea of Christianity or King Ashoka who supported the spread of Buddhism in Asia.

Sooner or later we will have to find a spiritual meaning in life, to assure value to the individual experience of being alive. We all should be educated to think for ourselves, free of indoctrinations on any kind. We should keep an open mind and study the history of philosophy— the most interesting entertainment around in these days of confusion.

I have deliberately opened myself to challenges, particular on the subject of transcendence. In the specifics, many statements made may turn out to be inaccurate, but that can be expected. It is equally true of science which is constantly revising and changing. Yet, the general principle of innate suggestibility would be difficult to disprove. This would require a convincing system. Single errors cannot bring it to collapse. Materialism stands, despite many errors, until another comprehensive system can fill the vacuum that is left before it collapses. Transcendence is confirmed by the wave structure of the phenomenal world. The hints of transcendental experiences means that knowledge is contingent on a serious desire of individuals to know. The hints are contained in the symbolism of dreams, as indirect information. But without prior saturation in a subject, and some knowledge of the interpretation of dream symbols, such information is lost and wasted.

The quality of human existence, which we perceive as participants from inside this value-creating process, has been steadily improving, partly through technology. There is a growing openness toward the idea of reincarnation. The possibility of transmigration of the soul can be noticed in the West, but nothing like a straight line that karma implies. Transmigration is more by choice and by attraction through suggestibility to values—disregarding time sequence and location, anywhere in the universe of mind.

The idea of a Universal Principle that repeats itself may first have been conceived by physicists. But the claim of universality was reduced to apply to only energy when a mathematical formula could not explain conscious orientation and the organization of reality into meaning. But this was overcome by claiming life and reality is meaningless. One may ask, why is everyone trying to hold on to life as long as possible?

The alchemists of the Middle Ages believed that an element of magic was involved in the world process. They obviously were correct. I also think there is a certain magic involved, namely that innate suggestibility can organize in ways we do not understand in detail. Perhaps other beings of older, more advanced planets have already developed suggestibility and use their emotions in ways which we fail to imagine on our present level.

Suggestibility is the greatest mystery of all, I personally believe. Behind the physical is the metaphysical, from where all the different ideas arise to proliferate and divide the world in to the multiplicity of beings and things that suggest a pluralistic universe. This is a universe, which according to evolution, is inspired by a single intent, expressed in one common principle, which repeats itself infinitely. One can conclude there may exist higher evolved beings, yet not perfect beings, only on different stages of evolution and technology from us. Even the continuity of the universe is based on imperfection, and all its creations are therefore imperfect. The value-creating process can never be completed because reason demands continuing imperfection to provide a persisting challenge to reorganize and change.

Psychology and psychiatry have found that emotions are an important part of reality. Mysticism and religion explain the whole process as phenomenon of spirit. The consensus of all perspectives is that stages and different modes suggest many manifest realities are possible, which to the normal experience of meaning, are tangible worlds.

As far as practical life experience is concerned, the physical and chemical explanations should be considered first. If these cannot explain what is taking place, metaphysical and spiritual explanations are necessary. This pattern is already applied in the field of health care. If the regular physician cannot cure an illness, then homeopathy and other suggestible-based methods should be used to challenge and shock the body to wake up and fight back. But to resort to using hypnosis means to interfere in the individuals value structure and is the last resort, because during a trance, suggestion can confuse the organization of the self and make it worse if one is not careful.

When meeting other persons, one does not exactly know what they think. This is good, but telepathic feelings provide some information, which could be right at that moment. But one must allow for ideas to change. It is therefore evident that suggestibility to value begins when we open our eyes to see others and think. If one is intimidated by the world and allows negative emotions to take root, then worry about the future can suggest itself. This is what existentialists mean by "angst." Apparently we think about the future and are aware of death—this is the negative side of intelligence. Ignorance is bliss, and worry about the future is spared to beings on less self-conscious levels. For them thinking is wasted of time. Only if one believes that physical dissolution of the organic body is the end of consciousness, angst becomes factor. Materialism implies finality which creates "existential angst." There may be massive evidence that the spirit survives, which means there is future after death, but one must be convinced which is difficult. The question is what awaits the spirit after death? Nobody knows. And the experience should be different for each individual. Certainly death is painful. One goes though life in the slow process of growth and must be content not knowing what will come as long it is something interesting. Concepts of the afterlife vary. The punishment for sins in everlasting hellfire is even worse than the nothingness of materialism. It is not surprising that Jung was outspoken against the idea of sin because it instilled guilt in suggestible persons. In fact, the majority of the mental ill in mental hospital hate themselves, believing they are wretched sinners fearful of punishment in hell. Such expectations of a future can drive some people insane if they really believe it.

If the transcendental seers of Tibet are right, our expectations guide us to a new beginning, in another experience by attraction to familiar values. This implies we are going where "we want to be" in the next life.

As a part of the universal consciousness, which never is not, we would not have to fear death, only transformation. And with a positive expectation, one should evolve higher, with a negative, lower. Innate suggestibility to value probably is the guide, and the adopted values in life set the direction. We don't know the specifics—all is speculation. If it is wishful thinking, I rather wish for an interesting future because the other alternatives are worse.

If one knew every detail of the future, than there is nothing new to look forward to experience. Exact knowledge of the future is undesirable. We are unable to know the future and unable to change it, except by intelligent judgement while we are living it could set the direction.

To expect exact knowledge in a constantly changing interactive process is unrealistic, because many values become obsolete. To repeat, nothing is constant but change. It is unwise to hold extreme views on anything, particularly about absolute justice. Because justice is a personal perception, it is impossible to fill. Even theologians have agreed in this point, and left justice to God in the next world. But God, if there is one, must forgive everything. The show must go on, individual actors are soon forgotten. This is how nature deals with the past, forgetting it. The future is where we are all going—we should think ahead. What is done cannot be undone, but one can learn something from it and avoid repeating the same situations in the future, unconsciously perhaps.

But from the recorded history of past events, one can speculate about the future. There may be some improvements in detail, but the principle itself remains unchanged. The same emotional impulses repeat themselves since we all are born ignorant and have to learn everything again from practical experience. The improvements that have occurred were paid for in blood and generally subconsciously raised individual awareness. Emotionally, we have become more tolerant and rational. Fanaticism and close-minded ignorance is natural to a primitive mentality. The expansion and refinement of consciousness, and empathy with other beings and suffering in general, is an indication that the being has moved to a higher mode—a positive sign of the value-creating process. But it does not alter the singular intention that the experience of interesting lives is the ultimate reason of being.

With the general rise of empathy we are reminded that we have evolved as socialized beings. Democratic constitutions will defuse intolerable conditions that lead to violent revolutions before they happen. The age of feudalism and its opposite, Communism, are not returning again in the same form. Humanity is further evolved to avoid explosive situations from discontent. We humans may be ready to understand the meaning of the world process as self-perpetuating value-creation, in which values attain meaning by their opposites. Equality and justice are not the issue in a process of differentiation. Alternative choices are pushing evolution forward.

Because the future is a logical progression of the past, some temporal regressions are possible. Since nobody seems in control, to prevent an ecological catastrophe, nature will have to regulate itself. The question is whether consciousness can exist without a biological infrastructure. This is a risky gamble mankind is taking. Perhaps clairvoyants such as Michel Nostradamus are correct, that we are allowing the destruction of the world, as we know it, to happen because the majority does not take the threat to the biosphere seriously. We have made social progress, but not because of a higher moral law in the human unconscious, only because organization requires cooperation and civilization is based on peaceful coexistence.

Normally people do not care what others have and do, as long as they can also live. Indeed, if life is an interesting experience, then material possessions can become chains that limit personal progress. Besides, one never could buy happiness. Happiness is an experience requiring tolerance. Materialism and mechanization are a wrong direction that lead to a dead end in evolutionary terms. We have now consciously entered into the information age. Through literature and the performing arts, we have further expanded our experiences of life. Through music, we have stimulated our emotions For some this is not enough—they want to experience illusions through chemicals, such as alcohol and drugs. But our brain can also produce its own stimulants. Foreign substances are actually destructive to organic systems. Our imagination is a better source of stimulation, but one can lose judgment of reality in either case. There is no need to rush; eternity is long.

If there is concern about human affairs on a higher level, which many intelligent persons assume there is, it is expressed in the form of suggestions and general guidance. The source is perhaps from departed humans who now exist on a higher level and who understand that one must evolve to maturity by personal experience. Herodotus, called this indirect guidance "providence." So did many others who have come after him.

Jung has called this guidance "a mysterious synchronisity of events," not only in the history of mankind, but also in the life of individuals. It is as if the meeting of individuals is sometimes arranged to bring them together. Jung also believed that a "collective unconscious dimension" is the source of transcendental dreams. Such ideas imply the existence of a spirit world. There are people who think that it is useful to communicate with spirits. But spiritualism also has a dark side, and one is advised to leave it alone until on gets there. That will happens soon enough.

Zen masters and yogis are trying to escape from the recurring cycles of birth and change—to eternal peace in nothingness. But is nothingness really bliss? Nothingness could actually be utter boredom and this is the reason for the existence of the universe. Obviously in nature everything tries to cling to life as long as possible. Evolution is the proof that the mainstream of life prefers to be something, however difficult as it may seem.

Let us look at existence from another perspective. Besides dreams, gifted psychics can provide information from departed spirits, since they can see and talk to the dead. Once I went to a spiritual meeting of a small group of persons, who had invited a psychic woman of Scottish descent to demonstrate her gift. This woman had never seen me before, nor has she since. I was only one individual in a group of about sixteen strangers. She translated messages to everyone in turn. When it came to me, she accurately described two recently departed persons, and the cause of my brother-in-law's death. The psychic was correct about general information. I had the feeling that the spirits knew what I wanted to know. She said, "There are two gentleman before me with messages for you. She described the first man by saying that he is telling me that he died from a pain in the stomach. I knew he died of cancer of the stomach. He wants to tell you "what you have started now will work out satisfactorily, and so will the long-range idea you pursue, but much later." And the other man, she described as shoemaker by trade, who to my knowledge was still alive. But she was right. A few days later I received a letter confirming he had recently died. This excludes telepathy or reading my mind. I had known this second man in my childhood days and since been out of direct contact with him. He also knew what I wanted to do, and said that it will need many more nails before it is completed. I was then concerned whether this book, which I then had just begun, would ever be completed. He was right. If nails mean thoughts, it took many nails at the rate I was going—more than thirty years of nailing thoughts down.

Skeptical persons will reject spiritualism, dreams and transcendence as useless and general, but in my case it encouraged me to continue when I lost hope that this book would ever be completed. The identification of these persons, confirmed by many who had something to do with psychics, strengthened my belief that death is not an end, and that other dimensions exist. I consider spiritual existence a level of human consciousness, nearer to what Jung describes as the collective unconscious of mankind. Yet, they cannot tell us anything reliable about "when and where," even if they wanted to do so. To know the future in detail would devaluate life. That is why such knowledge must be inconclusive and deliberately vague, and why reasonable persons prefer to find out by themselves what will happen, when it actually happens.

My intent is to understand the general principle of the universe by reason. I never expect to know all the many details which constantly change and make daily life interesting. Spiritualism and extrasensory knowledge has reached the limit placed on information long ago. It is only general, indirect information, pointing to a common principle. And if I now think about it, this principle has many elements of transcendental knowing in it. It is more transcendental, a vision of a new model of reality.

With the recent development of space flight, man is looking outward, beyond the human experience and must expect that consciousness is universal. We could assume that other species would have different values. Strange beings could exist, yet in order to have the ability to transcend, must have a selfless disposition, including loyalty to themselves. It is unlikely that a mechanical form of space flight could exist, traveling over such distances measured in light years, but the possibility should not be excluded completely. Since we already know from psychokinesis that to transcends into other modes is based on emotional synchronization, this means there are remote possibilities open for emotional space travel, but that is still gray territory.

Kirlian photography is an example that spirit survives physical amputation. Photographs show the images of severed limbs still in place although lost long ego. This implies there is a second spirit-body which survives the physical body, namely our ideal self. One could conclude that ideas and emotions can transcend the separations of levels, even if physical structures fall apart when the suggestion of value becomes too weak to keep it inspired to function.

From many facts a shift from a mechanical to an emotional, suggestible explanation of nature helps many humans make sense of the many exceptions that some individuals experience. The details of extrasensory experiences and dreams have many symbolic expressions to suggest. This is a logical conclusion. In a phenomenal reality, suggestibility is the essence of consciousness and explains the organization of waves into ideas and worlds, with one common, single purpose—to experience meaning through change.

That mechanization is the wrong direction leading away of natural evolution is obvious to reasonable persons with common sense, but materialism is still dominant in the world and in control of education, because it is supported by pragmatists. William James could not see a pattern in the phenomena of reality and concluded there is none. If something works and has usefulness that is enough reason to exploit it. But usefulness is not enough. Chemistry and atomic energy have catastrophic consequences on organic life, not only from single accidents like in Chernoble, a nuclear power station in Russia, but from the accumulation of radioactive materials and chemical waste over time. In fact, the accumulation of poisonous refuse of the technological civilization threatens our biological survival. The defenders of commercialism argue that pollution is a myth. The fact is that air and water are already poisoned. Obviously, it is not easy to change a system of such magnitude as scientifically based materialism, which promotes competition for material security for a life that is only short.

It is clear why the majority supports pragmatism; they simply cannot believe the world is a phenomenon of ideas. They take their senses as measure of reality, although we know that a process of synchronization of waves is creating the tangible world. If electronic devices such as television and holographs cannot convince people that waves can recreate life like phenomena, they would also not believe that invisible, electrical waves exist, nor would they believe the atomic bomb exists.

In spite of the reality of the wave nature of the world, skeptics have a compulsive loyalty to outdated concepts, accepting only physical evidence. Another unresolved issue lingering in the back of my mind is this: The presence of unnecessary brutality in nature—the reason for fear. Yet, if all is a phenomenon created by suggestion, then fear is also a suggestion. On the physical level, pain is real and meant to warn of injury. That is its only purpose. But without pain and suffering the world would not be taken for real. So, is there another way to learn except by experience? I think not.

Normally, we are concerned individuals and call ourselves civilized. But when we are emotionally involved, reason frequently fails us. This is a sobering throwback to our animal nature, particularly when each new generation is born intolerant. One can understand the pessimism of refined persons who know that the human race has not changed since the Stone Age, when the ethics of Darwinism—survival of the fittest—was hailed as intelligence.

Because there are occasional warnings to prepare us emotionally for the coming of tragic events, it appears that there is a concern for suffering on higher levels. But this concern does not take over and tell us how to live. We are meant to be free and can ignore suggestions—a right of self-conscious beings.

If we think that life is worth something, why do we allow incorrigible fanatics to start meaningless wars—wars that could affect all of mankind and devastate our blue planet? I am thinking of the War of Reformation in central Europe which ended only when the fanatics had killed themselves off and the survivors came to their senses. The survivors decided that they wanted to live and needed to rebuild their society or they would starve to death. After thirty years of religious hatred, the opposing sides agreed that religion should be a private choice, and not mandated by the state. Now, three hundred years later, many intelligent people are agnostic and would not go to war over personal beliefs. Most believe that there is something, like a neutral spiritual essence—a suggestible consciousness common to all beliefs that actually exists in all of us.

As if to answer my question as how to find an ending for this book, I had several symbolic dreams. I have already mentioned two dreams in this and other chapters.

Now at the very end of the book, let me mention another relevant dream, because reality is more than the physical world, it is in the mind that we live. I dreamt that I was employed at a bus stop, and a tourist bus has stopped for service. I was washing the windshield of the bus, which had a female driver at the wheel. She thanked me for my efforts, saying, "Now I can see clearer and further with a clean windshield." I went back into my little service station and, looking through the window saw this female driver standing there, talking to her flock of tourists. I thought she might want a refreshing drink and filled a glass with clear water and walked out to offer it to her. She drank it and said, "I needed the refreshment." Then she commented that usually she preferred a little sugar in her drinks, but plain water was still the best.

This dream answers many questions. For average expectations this philosophy may not be to the taste of some readers because they are loyal to their long-held values. The female driver is a condensed symbol of several ideas. By the symbolic washing of the window, I had made the driver and guide of people, who interprets the sights of the world, to see the future more clearly. Travelers along the road of life might prefer a sugary utopia to plain reality. But we have reached an impasse. The mechanization and digital developments lead to a dead end. It is necessary to see clearly and know in what direction we are driving or being driven.

The irony is that we believe that death is not the end and that there is a future. We cannot comprehend nothingness because our unconscious knows that consciousness is infinite. As physical organization we certainly are finite beings. Even if we call life a "finite experience within infinite consciousness," this leaves the existence of suggestible consciousness as absolute, as the first cause from which all else follows—a mystery.

Endnotes

1. *Psychic Explorations*
2. *Ibid.*
3. *Miracles and Other Realities*
4. *Parapsychology, the Controversial Science*
5. *Miracles and Other Realities*, and, *Parapsychology, the Controversial Science*
6. *The Centre of the Cyclone*
7. *Altered States of Consciousness*
8. *The Varieties of Psychedelic Experience*
9. *The Center of the Cyclone*
10. *Altered States of Consciousness*
11. *The Centre of the Cyclone*
12. *Ibid.*

Bibliography

Barrow, John D. *The Artful Universe.* Oxford: Claredon Press, 1995

Beaty, J. Kelly, Brian O'Leary, and Andrew Chaikin. *The New Solar System.* Cambridge, Mass: Sky Publish Corporation, 1981.

Broughton, Richard S. Parapsychology, *The Controversial Science.* New York: Ballantine Books, 1991.

Lilly, John. *The Centre of the Cyclone.* Princeton, N.J.: Juliet Press, 1972.

Masters, R.E.L., and Jean Houston, Ph.D. *The Varieties of Psychedelic Experience.* New York: Dell Publishing Co., 1966.

Mitchell, Edgar D., edited by John White. *Psychic Explorations.* New York: G.P. Putnam, 1974.

Morris, Richard. *The Nature of Reality.* New York: McGraw-Hill, 1987.

Pulos, Lee, and Garry Richman. *Miracles and Other Realities.* Vancouver and San Francisco: Omega Press, 1990.

Tart, Charles T., editor. *Altered States of Consciousness.* Garden City, N.Y.: Doubleday & Co., Inc., 1972.

Wyller, Arne A. *The Panetary Mind.* Aspen, Colo.: MacMurray & Beck, 1996.

About the Author

ERNEST (ERNST) J. LIEBSCHER was born June 1, 1925 in Northern Bohemia in a German-speaking area which was formerly Austria but in 1919 became Czecho-Slovacia. During a plebiscite, this area, called the Sudetenland, became a part of Germany in 1937. After high school, at age fifteen, Ernest took an apprenticeship in machine building. During this time he was exposed to many interesting people and expanded his education by reading philosophy and by taking language courses in English and French. In 1943, at eighteen, he was drafted into the German army. The war was already going badly and he experienced the collapse of Germany and the ensuing migration of population. His parents, who were considered ethnic Germans, were expelled from the new Czecho-Slovacia and joined the many homeless people of ravished Germany. After a few years, his parents settled near Frankfurt in the American sector. There Ernest tried to complete his formal education to become a mechanical engineer. After one and a half years, however, he gave up and decided to emigrate to Canada because he could not see a future for himself in Europe. In June 1951, Ernest arrived in Canada. Speaking fluent English, he found employment in toolmaking and eventually switched to precision measuring, mostly working in the mechanization of the auto industry.

He lives in the greater Toronto area in Ontario, Canada. In 1953 he married Christa-Anne. They have two sons and one daughter. Ernest retired at age sixty-three and spent the next eight years writing this book about the single principle of the universe which turned out to be anything but mechanical.